WORLD HERITAGE SITES OF AUSTRALIA

PETER VALENTINE
FOREWORD BY BOB BROWN

NLA PUBLISHING

CELEBRATING
NATIONAL LIBRARY
OF AUSTRALIA
PUBLISHING
50 YEARS

CONTENTS

First Nations peoples are advised this book contains depictions and names of deceased people, and content that may be considered culturally sensitive.

Sunset at a desert in Mungo National Park, New South Wales.

TRADITIONAL CUSTODIANS

This book was written on Ngadjon Country and produced on Ngunnawal and Ngambri Country, with valuable input from Traditional Custodians around Australia. The author and the National Library of Australia acknowledge Australia's First Nations peoples, the First Australians, as the Traditional Owners and Custodians of this land. We give respect to the Elders, past and present, and through them to all Australian Aboriginal and Torres Strait Islander people. We thank the Traditional Custodians of Australia's World Heritage sites for their continuing care of Country, including:

World Heritage Site	Traditional Custodians
Kakadu National Park	Bininj/Mungguy
Great Barrier Reef	Erubam, Ugarem and Meriam Le, Kaurareg, Gudang, Yadhaigana, Wuthathi, Kuuku Ya'u, Kanthanumpun, Uutaalgnunu (Night Island), Umpila, Angkum, Lama Lama, Paal Paal, Guugu Yimithirr Warra, Ngulan, Yuku Baja Muliku, Eastern Kuku Yalanji, Wanyurr Majay, Yirrganydji, Gimuy Yidinji, Gurabana Gunggandji, Guru Gulu Gunggandji, Mandingalbay Yidinji, Lower Coastal Yidinji, Mamu, Djiru, Gulngay, Girramay, Bandjin, Warrgamay, Nywaigi, Manbarra, Wulgurukaba, Bindal, Juru, Gia, Ngaro, Yuibera, Dharumbal, Woppaburra, Taribelang Bunda, Bailai, Gooreng Gooreng, and Gurang
Willandra Lakes Region	Paakantji, Mutthi Mutthi and Ngiyampaa
Tasmanian Wilderness	palawa
Gondwana Rainforests of Australia	Anaiwan, Banbai, Birpai, Bundjalung, Thunggutti, Geawegal, Githabul, Gumbaynggirr, Jukumbal, Ngarrabul, Wangerriburra, Widjabul Wia-bul, Wonaruah, Worimi, Yaegl, Yugambeh Ugarapul, and Yuggera
Uluṟu-Kata Tjuṯa National Park	Yankunytjatjara and Pitjantjatjara
Wet Tropics of Queensland	Eastern Kuku Yalanji, Western Yalanji, Koko Muluridji, Yirrganydji, Djabugay, Yidinji, Gunggandji, Mbarbaram, Ngadjon-Jii, Mamu, Jirrbal, Gulngay, Djiru, Girramay, Warrungu, Bandjin, Gugu Badhun, Warrgamay. Nwaigi and Wulgurukaba
Shark Bay, Western Australia	Malgana, Nhanda, Yamatji and Yingkarta peoples
K'gari (Fraser Island)	Butchulla
Australian Fossil Mammal Sites	Riversleigh: Waanyi; Naracoorte: Meintangk
Greater Blue Mountains Area	Darkinjung, Dharawal, Dharug, Gundungurra, Wonnarua and Wiradjuri
Purnululu National Park	Gija and Jaru
Royal Exhibition Building and Carlton Gardens	Kulin Nation, Wurundjeri people
Sydney Opera House	Gadigal people of the Eora Nation
Australian Convict Sites	Dharug Nation, Burramattagul people; Gadigal people of the Eora Nation, palawa, Panninhe, Puthikwilayti, Muwinina, Pydairrerme peoples and Whadjuk people, Noongar
Ningaloo Coast	Yinggarda, Baiyungu and Thalanyji
Budj Bim Cultural Landscape	Gunditjmara

FOREWORD

World Heritage is the pride of our planet, the status accorded to the most astonishing natural and cultural sites on Earth. In this second edition of *World Heritage Sites of Australia*—featuring a new chapter on the incredible Budj Bim Cultural Landscape—Professor Peter Valentine presents the places in our nation that have qualified as World Heritage, alongside the Grand Canyon, the Serengeti, the Great Wall of China and Westminster Abbey. He outlines the unique values of the 20 Australian sites and gives us his personal appreciation of each place, accompanied with brilliant photographs.

Looking out over the Kakadu wetlands, Professor Valentine describes the vibrant tropical landscape and the rich Aboriginal artworks adorning its rock galleries as entwined and inseparable. He quotes Gagudju Elder Bill Neidji who encapsulated humanity's relationship with the natural planet: we are all one.

From Heard and McDonald Islands, which are almost totally natural, to the Tasmanian Wilderness, where human and natural heritage both excel, to the Melbourne Exhibition Building and Carlton Gardens, where the grandeur of modern human creativity and construction predominate, here is a compendium of Australia's greatest treasures.

A strong motivation for establishing the World Heritage Convention was global outrage that the statues of the 3,000-year-old temple of Abu Simbel were to be drowned by a 1960s dam being built across the Nile River in Egypt. An international effort rescued the statues by carving them out of their cliff face and removing them to higher ground.

In Australia, there was a parallel furore over the loss of beautiful Lake Pedder and its extensive highland beach, in southwest Tasmania, also to be drowned by the dams of modern progress. Alas, there was no rescue here. The national government lacked the constitutional power to intervene on the Tasmanian state government's project. The Lake Pedder tragedy added impulse to Australia becoming one of the first nations to sign the World Heritage Convention, in 1974, and to our first three properties, the Great Barrier Reef, Willandra Lakes and Kakadu, being listed as World Heritage in 1981.

This 'external treaty' empowered the Australian Government to overrule the Tasmanian government a decade after the loss of Lake Pedder and stop the Franklin Dam in Tasmania, saving the Franklin River along with much of the rest of wild Tasmania, after it was listed as World Heritage in 1982. Similarly, it empowered Canberra to intervene in Queensland and save the Wet Tropics (Daintree) rainforests from piecemeal destruction in 1988.

Half a century after the World Heritage Convention came into being, there are still points of contention over both listed and unlisted sites in Australia. Global warming and ocean acidification and pollution are threatening the Great Barrier Reef. There is mounting concern about unlisted but World Heritage-worthy sites like Tasmania's threatened takayna/Tarkine rainforests and the world's greatest rock art realm, Murujuga, in Western Australia, targeted for more industrial development even though the state government is committed to its World Heritage nomination. And how about the absence from the World Heritage list of the greatest wilderness of all, Antarctica?

This book is an uplifting introduction to our grandest natural and cultural heritage and a challenging reminder that its protection has not always come easily. There is much yet to be done to both guarantee the safety of listed sites and win listing for the nation's remaining globally significant jewels that too often remain open to theft and despoliation. This book inspires the hope that they too will be listed and protected in the near future.

Thank you, Peter Valentine!

Dr Bob Brown

Bob Brown at the National Library of Australia in 2017.

AUSTRALIAN WORLD HERITAGE SITES SUMMARY

World Heritage Site	Location	Listing	Type
Kakadu National Park	Northern Territory	1981	Mixed
Great Barrier Reef	Queensland	1981	Natural
Willandra Lakes Region	New South Wales	1981	Mixed
Lord Howe Island Group	New South Wales	1982	Natural
Tasmanian Wilderness	Tasmania	1982	Mixed
Gondwana Rainforests of Australia	New South Wales, Queensland	1986	Natural
Uluṟu-Kata Tjuṯa National Park	Northern Territory	1987	Mixed
Wet Tropics of Queensland	Queensland	1988	Natural
Shark Bay, Western Australia	Western Australia	1991	Natural
K'gari (Fraser Island)	Queensland	1992	Natural
Australian Fossil Mammal Sites	Queensland, South Australia	1994	Natural
Macquarie Island	Tasmania	1997	Natural
Heard and McDonald Islands	Australian Territory	1997	Natural
Greater Blue Mountains Area	New South Wales	2000	Natural
Purnululu National Park	Western Australia	2003	Natural
Royal Exhibition Building and Carlton Gardens	Victoria	2004	Cultural
Sydney Opera House	New South Wales	2007	Cultural
Australian Convict Sites	Tasmania, New South Wales, Western Australia, Norfolk Island	2010	Cultural
Ningaloo Coast	Western Australia	2011	Natural
Budj Bim Cultural Landscape	Victoria	2019	Cultural

INTRODUCTION TO WORLD HERITAGE

One pathway to the establishment of the World Heritage Convention began with ancient Nubian monuments at Abu Simbel, part of Egyptian heritage from thousands of years ago, and a decision to build the Aswan Dam across the River Nile. Facing the potential destruction of this exceptional legacy, the governments of Sudan and Egypt asked the United Nations Educational, Scientific and Cultural Organization (UNESCO) to help their attempt to save at least some of the history that was threatened by the dam. In 1959, UNESCO launched an international campaign seeking cooperation and financial support to relocate the temples urgently, ahead of the rising waters. Some 50 countries cooperated, with funds and expertise. This amazing project succeeded in preserving ancient Egyptian culture by dismantling, relocating to dry land and reassembling these huge temples and statues exactly as they were. It demonstrated that international cooperation for heritage protection could achieve great outcomes. More collaboration followed, including the development of a formal convention with the direct involvement of the International Council on Monuments and Sites (ICOMOS).

A second path emerged from the national park movement. At the White House Conference on Natural Beauty in 1965, it was proposed that some natural heritage was of such significance that there should be a means of stimulating international cooperation to protect it. The International Union for Conservation of Nature (IUCN) developed ideas about this through its membership, eventually taking a draft agreement to the 1972 United Nations (UN) Conference on the Human Environment in Stockholm. There, a single text was agreed upon, linking natural and cultural heritage. This text was adopted by the General Conference of UNESCO on 16 November 1972. The Convention Concerning the Protection of the World Cultural and Natural Heritage, known as the World Heritage Convention, had come to fruition.

The World Heritage Convention was one of several multinational environmental agreements that emerged during the 1970s, a period when the world became increasingly concerned for the future of the planet and its peoples, perhaps stimulated by the first photographs of the earth seen from space. Some early thinking around World Heritage gave a strong focus to the celebratory context. Dr Jim Thorsell, the long-time senior advisor on World Heritage for the IUCN, would often refer to the successful listing of a natural heritage site as 'the Nobel Prize for nature' or to the list as 'nature's hall of fame'. Others would write about the 'best of the best' to draw attention to the exceptional qualities of places on the World Heritage List. For some, the listing was a badge of merit—international recognition of a site's quality. At the time of writing, there are 1,157 such sites around the world.

Many World Heritage areas are visually splendid and iconic of the country in which they feature. The Taj Mahal of India needs no introduction; this 'jewel of Muslim art' was listed in 1983. The rock islands of Palau have been visited by very few people, but aerial views of their natural beauty were well recognised before their listing in 2012. Similarly, Australia's Great Barrier Reef was a global icon long before it was listed, as were Uluṟu, Kakadu and the Sydney Opera House. Exploring the World Heritage List country by country often creates a little surprise, as we recognise places we have never visited. In the United States of America, we note the famous Statue of Liberty and the Grand Canyon; in France the Eiffel Tower, part of the Paris, Banks of the Seine site. The distinctive forms of Stonehenge are easily recognisable, as is the Great Wall of China. But many World Heritage areas are not so well known or have attributes that are not easily discerned. Examples include the Messel Pit Fossil Site in Germany that allows us to understand much about life on earth 48 million years ago; Dinosaur Provincial Park

Native pigface in the Tarkine Wilderness, Tasmania.

World Heritage has been a significant part of my life since 1989, when I first became a member of the IUCN World Commission on Protected Areas. Since then, I have been involved in the evolving operation of the World Heritage Convention across many countries, both in a technical sense (assessment, advice and training) and in enjoying the amazing variety of our global heritage. I have visited over 50 World Heritage sites in 20 countries and few experiences have given me greater pleasure.

in Canada, with its rich array of dinosaur remains from 75 million years ago; and the Willandra Lakes Region, where an understanding of climatic oscillations, combined with the archaeological evidence of human occupation 45,000 to 60,000 years ago, led to the dramatic revision of previous estimates of Indigenous occupation in Australia.

Over time, the emphasis has shifted from identification of World Heritage sites to protection of them, and the World Heritage Committee has clarified signatories' obligations under the convention. By signing the convention, the UNESCO website notes, 'each country pledges to conserve not only the World Heritage sites situated on its territory, but also to protect its national heritage'. Signatories agree to cooperate with the international community in meeting obligations and to take a number of specific actions. Article 4 of the convention spells out the duties of each state party (country) in the 'identification, protection, conservation, presentation and transmission to future generations of the cultural and natural heritage' and specifies that each country must do all it can 'to the utmost of its own resources' to this end. Although signing on to the convention is voluntary, once committed, a country takes on a series of responsibilities. Over the years, the World Heritage Committee has developed a set of operational guidelines that assist countries to better understand their responsibilities (see a list of these on page 290). The current version of this technical document is 188 pages long, covers a vast range of topics, and is periodically rewritten. The convention itself is just 16 pages long and has never been revised.

As well as having outstanding universal value (see criteria on page 290), each World Heritage site must meet certain conditions of integrity and authenticity. These are assessed at the time of inscription and are regularly monitored to ensure that the property is in good condition. 'Integrity' relates to the property's wholeness or intactness, with implications for the area of the property, the precise boundaries identified, the adjacent land use and other factors that may inhibit its continuing survival. In natural heritage sites, this often involves a good knowledge of biological or geological processes that affect the site and its condition. The IUCN assesses integrity as part of its evaluation on behalf of the committee when a site is nominated. For cultural heritage sites, integrity involves the physical structure of the property and a range of factors that may affect its condition.

'Authenticity' refers particularly to cultural heritage and the credibility of claims made about a site in nomination documents. ICOMOS applies a range of practical techniques to develop the advice it provides to the World Heritage Committee on each nomination, including consulting with experts from around the world on the various attributes of each site.

The operational guidelines note that the 'convention is not intended to ensure the protection of all properties of great interest, importance or value but only a select list of the most outstanding of these from an international viewpoint'. Such properties are recognised by the World Heritage Committee on the basis of being able to meet specific criteria for natural and cultural heritage. These criteria are occasionally refined. At least one of these ten criteria must be met by a property before the committee will inscribe it on the World Heritage List. Some sites are mixed, having elements of both natural and cultural heritage of outstanding universal value.

It is solely the committee that determines whether a property has outstanding universal value and therefore merits inscription on the list. Once the committee receives a nomination, the advisory bodies IUCN and ICOMOS undertake careful evaluations of the site. Scientists and technical experts from around the world will usually make one or more site visits as well as desktop reviews. These experts then report on their findings, helping the committee reach its decision.

The rising and setting sun illuminates the vivid red rock of Uluṟu.

The first properties to be inscribed on the list in 1978 included such obvious gems as the Galápagos Islands (Ecuador), Yellowstone National Park (United States of America), the Historic Centre of Kraków (Poland) and the medieval Rock-hewn Churches, Lalibela (Ethiopia). The Nubian Monuments from Abu Simbel to Philae (Egypt), which had been instrumental in the development of the idea of World Heritage, were added to the list in 1979, along with many other sites across the world. The World Heritage Convention is one of the most popular environmental conventions, with 194 countries joining (only three UN members have not: Liechtenstein, Nauru and Tuvalu).

In addition to the World Heritage List, the committee maintains a second list: the List of World Heritage in Danger. The committee may add a property to the 'in danger' list if the property is threatened by a serious and specific danger. One intention is to generate support from the international community to help remove the threats. When the committee recognises that the threat has been removed, it will take the property off the 'in danger' list.

Statues of Ramses II line the main temple at Abu Simbel in Aswan, southern Egypt. The temple was carved out of the mountain between 1274 and 1244 BCE.

When Yellowstone National Park was threatened by a proposed adjacent mining project, the site was added to the List of World Heritage in Danger. Thanks to a series of legal challenges, the proposed mine was cancelled and a number of tourism management issues addressed. The success of these measures allowed the committee to remove Yellowstone from the 'in danger' list. There are many other examples of sites going onto the List of World Heritage in Danger for a period and then being removed once threats are averted. On occasion, a property may go straight to the 'in danger' list at the same time as it is inscribed as a World Heritage site. This may happen when war or civil unrest force an emergency listing, as in Afghanistan's Bamiyan Valley after the Taliban deliberately destroyed two Buddha statues. The World Heritage Committee inscribed the site in 2003 and immediately added it to the 'in danger' list, where it remains today.

On occasion, a state party makes decisions that render the heritage lost and at that time the committee may decide to remove the place from the World Heritage List. The first site removed was Oman's Arabian Oryx Sanctuary, in 2007. It was inscribed in 1994, primarily for its habitat for endangered bird and mammal species. When Oman reduced the size of the reserve by 90 per cent and started a program of hydrocarbon extraction over the former habitat that plunged the population of Arabian Oryx from 450 to 65, the committee determined that the site had lost its outstanding universal value. It expressed regret that Oman had failed to fulfil its obligations and for the resulting loss of heritage. Germany's Dresden Elbe Valley is another example. Listed as a cultural landscape in 2004, it was threatened by plans to build a major four-lane bridge through its heart and removed from the list by the committee in 2009. This is a case where one arm of government made a decision with adverse effects for another.

Australia was one of the first countries to join the World Heritage Convention—in 1974—and currently has 20 World Heritage areas, with sites in every state and territory except the Australian Capital Territory. In many cases, nominations for World Heritage in Australia followed, or were part of, political disputes and have led to legal challenges, community conflict and attempted interference in the operation of the World Heritage Convention. Despite this, it is clear that the convention has effectively increased recognition of Australia's natural and cultural heritage and informed improved management of the listed sites.

The managers of these wonderful places bring passion and commitment to the ideals of the convention and combine the needs of protection with the desire to share these treasures with the rest of the world. It is the work of these frontline rangers and managers, and the communities that support them, that will be critical in determining the long-term survival of our global heritage. The World Heritage Convention is a perfect example of the thoughtful injunction to 'think globally, act locally'. The best prospects for the future rely on a well-informed and active community.

KAKADU NATIONAL PARK

Long ago, a Bininj/Mungguy man stood under the overhang of a sandstone outcrop. He inspected the rock wall where he planned to make his mark. He had ground ochre into a fine powder and mixed it with water to form a slurry. His paintbrush was a combination of chewed twig and hair, perfect for applying the vivid red colour to the underlying white foundation. He knew the subject intimately; it was an important food. He visualised the complete animal: its outline shape, its interior with strong backbone, finer lateral bones and the different internal organs. Taking up the brush, he stroked the outline then continued the work. Each stroke brought out the living animal, the spirit of being. Like every artist, he knew just where in the rock the *namarngorl* resided and his task was to reveal it with careful, precise work. Slowly the internal features emerged: the distinctive backbone holding it together, the bulging stomach and connecting tissue, the gill flap and the jaws. All were framed with fins and tail, illuminating its identity: barramundi. He returned many times to complete this painting and, over many generations, his descendants kept the tradition going. Today, we call this the x-ray style.

THE MOST REMARKABLE ELEMENT of Kakadu National Park is the way outstanding natural heritage is intimately connected with exceptional cultural heritage. It is this juxtaposition of intact natural features and ongoing culture that most impressed the UNESCO evaluators in 1981 and that today inspires and astounds visitors.

Fifty to sixty-five thousand years ago, the first people were already living at Kakadu in the far north of Australia. This was a time when the great ice sheets of the latest glacial period were advancing south over much of Europe. It would be another 30,000 years before the glaciers reached their maximum extent and sea level was at its lowest. The ancestors of the first people had made their way from Africa across the edge of the great continent of Eurasia and eventually through south-east Asia and into Australia. They were the vanguard of other later journeys out of Africa and their migration took them far across the world. We do not know over what period of time this journey took place.

What these first people found at Kakadu was a mix of high sandstone plateaux and minor river systems in a grassy savanna woodland. The area had not yet developed the rich wetlands of today; they would follow much later, during

PREVIOUS PAGES: Turtle and barramundi, Nanguluwurr rock art site.

the course of settlement by these first Australians. The fauna they encountered bore no relationship to that left behind in Asia and Africa. Australia is a completely different biogeographic realm and, when the first people arrived, everything was strange and new to them. Apart from the novelty of marsupials like kangaroos, there were also giant herbivores (diprotodons) and large carnivores (marsupial lions), which were widespread across the continent. Today, these wondrous animals can only be imagined.

Over the subsequent tens of thousands of years, Kakadu people came to know the land intimately. They have conveyed their knowledge of it across generations through song, stories, dance and art; theirs is among the most remarkable art in the world. Rock art and archaeological evidence record social and ritual activities associated with the hunting and gathering traditions of Indigenous people over at least 50,000 years. They also give insight into how a community changed and adapted to a very long period of climate change.

The earliest art of handprints and grass prints (including a sorghum species still prevalent in the area) dates back at least 35,000 years. Handprints were created by dipping a

ABOVE: The Arnhem Plateau, surrounded by Kakadu's monsoon forest.

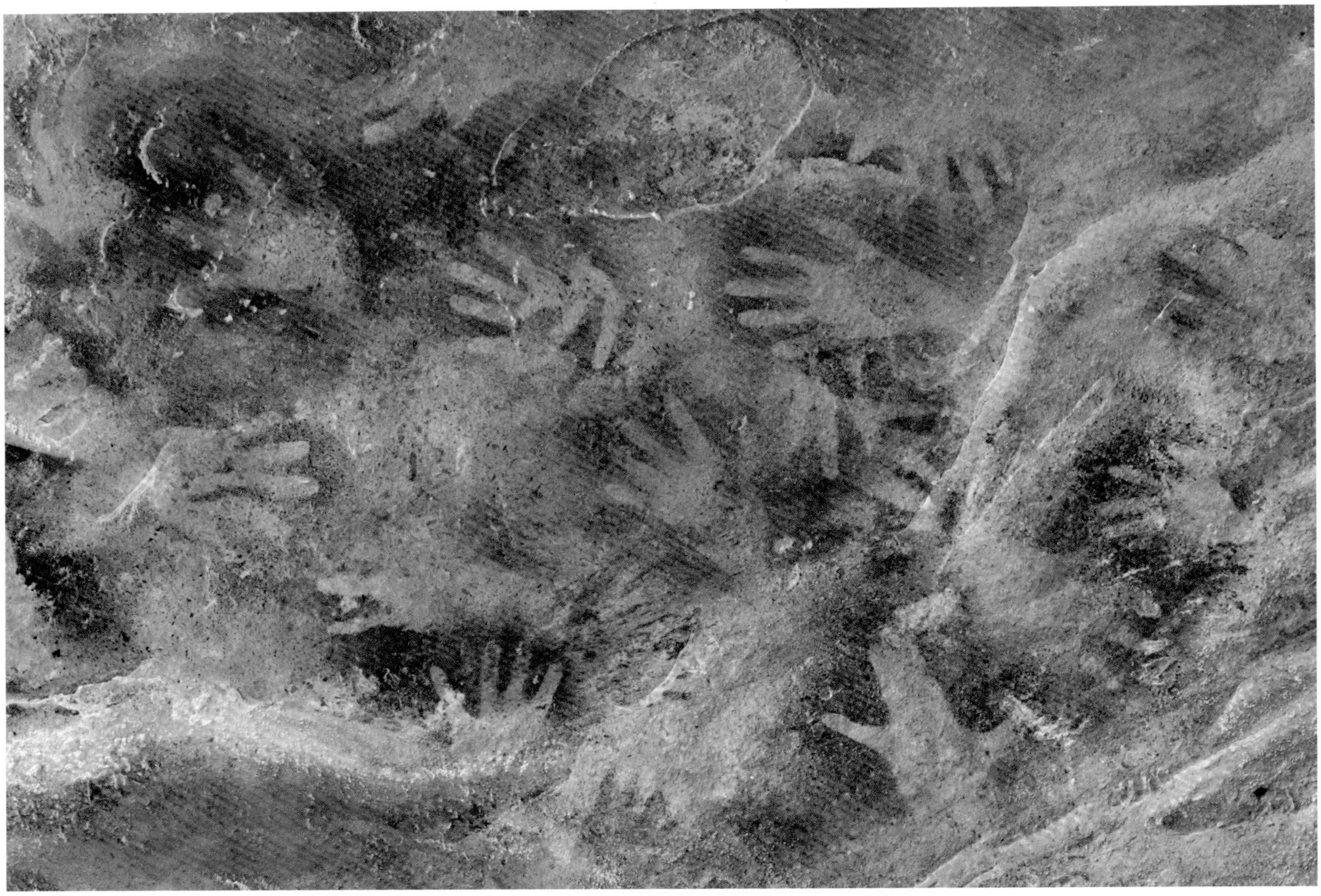

hand into some liquefied ochre and pressing it against an exposed sandstone wall, usually under an overhang or in a cave. Such handprints were often arranged in a pattern or cluster. During this period, Kakadu was well inland from the coast due to the much lower sea level. Some 20,000 years ago, at the time of the last glacial maximum when sea levels were at their lowest and glaciers their biggest, the art became more naturalistic and included animals and humans drawn in a way that enabled species recognition. Over the following 10,000 years, the art styles evolved from stick-shaped humans and other animals to more stylised figures embedded with richer symbolism. Hand stencils also emerged, created by blowing a mouthful of liquid ochre against a hand held to the rock surface.

By the time sea levels reached those of the present, and estuarine environments had reached Kakadu, the distinctive x-ray form of art had emerged. X-ray art shows the internal organs and bones of the animal. The subject is initially outlined in white and then the internal detail is rendered in other colours, usually red. Most recently, over the past 5,000 to 10,000 years, the art reveals the dominance of freshwater animals, including magpie geese. It also records contact between the Indigenous people and the Macassan people from Celebes (Sulawesi, Indonesia) and, more recently, with Europeans. This enormous sequence of art, the longest and greatest ever known, is well preserved at over 5,000 sites throughout Kakadu National Park and is a major reason why the area was inscribed on the World Heritage List. The art is so extensive and comprehensive that archaeologists have been able to interpret a great deal about life in the area over many thousands of years.

Much of the art represents animals of great interest to people, including kangaroos, crocodiles, fish species, crayfish, birds, reptiles and echidnas. Many species are easily recognised, such as the impressive black-necked stork that fishes the edge of the wetlands and the magpie goose, a significant food source. There are also some surprising depictions, including the now-extinct thylacine. While present across the tropical north tens of thousands of years ago, the thylacine subsequently retreated south and was eventually made extinct by Europeans in its final stronghold of Tasmania. Similarly, rock art depicting a numbat gives evidence of the much wider distribution of that species compared with its current habitat in a tiny area of south-western Australia, where it is on the verge of extinction.

THE ART IS SO EXTENSIVE AND COMPREHENSIVE THAT ARCHAEOLOGISTS HAVE BEEN ABLE TO INTERPRET A GREAT DEAL ABOUT LIFE IN THE AREA OVER MANY THOUSANDS OF YEARS.

The landscape of around 5,000 years ago was much more diverse than that found by the first Aboriginal settlers. It included coastal and estuarine environments (rich in fish and other wildlife), forests of eucalypt and paperbark, and floodplain areas that cover much of the land during the wet season and provide refuges in the dry season. The climate was wetter and contributed to the development of extensive river systems in Kakadu, with associated wetlands. The core of the park is the sandstone plateau that, at 300 metres high, dominates the eastern and southern parts of Kakadu. Over millennia, this has been eroded and undercut by water, creating spectacular waterfalls in some places, as well as ravines, waterholes and streams that feed the meandering rivers and associated tropical lagoons. The exposed escarpment is beautifully coloured in reds, browns and greys, varying greatly in response to the time of day and the available sunlight. These outcrops provide the many art galleries that have been used for tens of thousands of years.

One of the astonishing things about Kakadu—and a core reason for its listing as a World Heritage area—is that it

Nanguluwurr rock art site, showing human hand shapes and evidence of early contact.

is a living cultural landscape within a huge area of natural ecosystems that have hardly been affected by Europeans. It is one of the world's largest protected tropical areas—almost two million hectares in area—and includes the entire South Alligator River system within its boundary. Such an extensive area of intact ecosystems is rare in itself across the world, but its existence alongside an ongoing cultural presence is exceptional.

The descendants of the first people are still active in maintaining traditions and managing their Country. The name Kakadu was popularised by Sir Walter Baldwin Spencer, an ethnologist who worked in the area in the early twentieth century. It was his own rendering of the name for the main language spoken in Kakadu at the time, which was Gagudju. Sadly, this language is now extinct. The people of Kakadu were made up of many clans and had no collective name for the wider community. When forced to deal with European laws and ways, the local clans initially formed an alliance called the Gagudju Association, but more recently there have been a number of associations representing the local people. Bininj/Mungguy is a term used to refer to the local Aboriginal people of the park. Bininj is a Kunwinjku and Kundjeymi word from the north of the park and Mungguy is a Jawoyn word for man. Many Bininj/Mungguy people live and work in the park today and their knowledge and activities are essential for the maintenance of its natural and cultural heritage.

The knowledge system of Indigenous custodians was developed over thousands of years and includes details of the ecosystem and all its component parts. The essential philosophy of Bininj/Mungguy is that humans are part of the environment and are responsible for nurturing and conserving all of nature. The strong connection between the material and spiritual worlds gives rise to understanding and guides action for Bininj/Mungguy. The creator beings established protocol and practices, and the supporting stories are often a focus for the artwork we see today.

It was the earth mother, Warramurrungundji, who created the land and the people at the beginning of time. She also planted the crucial food plants and other resources so her people could survive. Many other creator beings emerged to give guidance and knowledge about survival and to make different parts of the landscape.

This connectivity between humans and all other life is succinctly expressed by the late Bill Neidjie, an Elder of great standing and possibly the last speaker of Gagudju:

Animals like family to us,
Earth our mother,
Eagle our cousin,
Tree is pumping blood like us,
We all one.

Many ancestral beings can be seen in the rock art. At Burrungkuy (Nourlangie Rock) there is a painting of Namarrkun, the lightning man, who brings the spectacular thunder and lightning displays in the lead-up to the wet season. Not all the ancestral beings were good and there are also stories of the bad spirits.

OPPOSITE: Jim Jim Falls, the tallest waterfall in Kakadu, descends from the escarpment to the forest below.

ABOVE: Kakadu Elder Bill Neidjie.

As might be expected from a culture that has existed for thousands of years, the Indigenous community has an intimate knowledge of the land that is often much more comprehensive than that of Europeans. A good example is the way in which seasons are marked. For most non-Indigenous people, the Kakadu seasons are simple—the wet (from about December to March or April) and the dry (May to November). A much more complex appreciation is reflected in the six seasons of the Bininj/Mungguy calendar.

English Name	**Bininj/Mungguy Name**	**Months**
Monsoon Season	*Gudjewg*	December to March
Harvest Time	*Banggerreng*	March to May
Cool Weather Time	*Yegge*	May to June
Early Dry Season	*Wurrgeng*	June to August
Hot Dry Season	*Gurrung*	August to October
Pre-Monsoon Season	*Gunumeleng*	October to December

Each of the Bininj/Mungguy seasons is associated with the transforming landscape and a series of ecological events and harvesting opportunities. For example, in *gurrung* the freshwater crocodile and the pig-nosed turtle lay eggs; in *yegge* the estuarine crocodiles move from the wetlands back into the main rivers. In *banggerreng* the magpie geese nest, a major event in the Bininj/Mungguy calendar.

It is possible today to stand on one of the rocky sandstone outliers, gaze across the extensive wetlands to distant escarpments and see huge flocks of magpie geese and other waterfowl flying between their daytime roosts and feeding areas. At the base of the rocky escarpments, along the many streams flowing out of the sandstone, are areas of monsoon forest with their particular plants, birds and other animals. Across the intervening country are grasslands and woodlands dissected by the gradually merging streams and rivers heading to the coast. Within this country are the many billabongs, lagoons and paperbark swamps that support Kakadu's richest wildlife. The wetlands are themselves recognised as being of international significance, containing outstanding aggregations of waterfowl. On occasion, millions of waterfowl gather here—a spectacle rarely seen anywhere else in the world.

Kakadu is a major staging post for the many migratory species from the northern hemisphere that spend the warmer months in Australia. Over 30 species of wader have been recorded within the wetlands and floodplains. This is one factor that led to Kakadu's recognition under the Ramsar Convention as a wetland of international significance. Among other species, sandpipers, godwits and stints take up residence on the shoreline and floodplains during the summer months, arriving after flying thousands of kilometres on their hazardous annual journey from arctic and subarctic regions.

The richly diverse mosaic of ecosystems that make up Kakadu provides for an astonishing diversity and abundance of wildlife. There are at least 2,000 plant species recorded in the park; some species' habitats extend across the tropics while others are confined to the Arnhem Plateau. Almost 80 mammal species live in the area (one-quarter of Australia's land mammals), alongside 280 bird species. Reptile species number 132, including both estuarine and freshwater crocodiles, the fascinating and large pig-nosed turtle and many snake and lizard species. Within the

aquatic environments are 27 frog species and over 300 fish species, including many depicted in the rock art, such as barramundi, mullet and catfish. An outstanding example of these fish paintings—in the x-ray style—can be seen at Ubirr, a spectacular series of galleries open to the public. Less well known are the 55 species of termite and their extensive sets of mounds, which are so obvious in parts of the landscape. Over 200 species of ant, representing ten per cent of the world's species, are also found in this World Heritage area.

The stone country of Kakadu, ancient sandstone rocks that make up the Arnhem Plateau and the hundreds of kilometres of escarpment, provides an imposing focus for the area. These rocks were laid down 1,650 million years ago, making them among the oldest sedimentary rocks on earth. Erosion has produced numerous gorges in the plateau. The crevices that have formed along the escarpment create waterholes and streams that channel seasonal water flows out onto the adjacent plains. The Kombolgie Formation is remarkable in that it has remained

A flock of magpie geese takes to the air over the Kakadu wetlands.

relatively stable for an extraordinary length of time. There has been no significant disruption of the sediments since formation, and exposed faces are in almost the same condition as when they were deposited. Stone country is found mainly in the eastern part of Kakadu and beyond into Arnhem Land, forming the upper catchment of the South Alligator River that bisects the park from south to north before discharging into Van Diemen Gulf.

The sandstone ecosystems of the Arnhem Plateau are home to many endemic species, including the banded fruit-dove, a shy bird that is most often seen feeding on figs along the edges of the escarpment or its outliers. The black wallaroo stays mostly within the rocky plateau, but in the dry season may move down along the edge to find fresh grass; its habitat is confined to this area. The rare white-throated grasswren requires old spinifex country and lives in areas of the plateau that have not been burned for a long time. Pairs or small parties of the endemic chestnut-quilled rock-pigeon live on the rocky ledges. Many frogs make use of the rock pools and at night a number of gecko species can be found. Mammals in this area are mostly nocturnal, including the rock ringtail possum and the sandstone antechinus, a tiny marsupial carnivore. The large and elusive ghost bat roosts and breeds in some of the caves and these rocky areas also provide prime habitat for the predatory rock python. Two different species of rock wallaby use this habitat. The white-lined honeyeater, another endemic bird, inhabits the sandstone country and its call can be heard throughout, especially near waterholes. It is partly this great diversity of wildlife that led to World Heritage listing.

The northern and western parts of Kakadu are dominated by tropical woodlands (savanna) and the floodplains of the four rivers that flow north—the East Alligator, South Alligator, West Alligator and Wildman rivers. The floodplains and the rivers and billabongs make up the wetlands of international significance, home to a vast variety of aquatic wildlife, including birds of astonishing abundance. These are the strongholds of huge estuarine crocodiles, which feed on the abundant fish and other animals that share their habitat. Many distinctive bird species live in the tropical savanna, including beautiful Gouldian finches, masked finches and long-tailed finches. Other seed-eaters such as partridge pigeons and brown quail also breed here after the wet. During the main rainy season, the rivers and streams flood out over the plains and reconnect all the waterholes and billabongs in a continuous sheet of water, allowing fish and other animals to migrate across the plains. Adjacent to these wetlands, on areas of poor drainage, are paperbark swamps, shallow but rich with plants and animals. In the dry season these dry out completely and so wildlife have developed various strategies to survive until the next wet. Giant water lilies produce seeds that lodge in the mud at the bottom of the drying swamp; long-necked turtles prepare a compartment beneath the mud in which to aestivate. Once the mud dries they are sealed inside the hard clay with just a small airhole through which to breathe until the rains come to release them into a new-season lagoon.

Another of Kakadu's distinctive ecosystems is the semi-deciduous monsoon forest. It frequently occurs as small patches where streams emerge from the sandstone plateau. Much further downstream, the forest appears as riparian strips along the rivers' more permanent freshwater sections. In the wet season, these are rich with fruit and therefore with fruit-eating birds and mammals: fruit-doves, orioles, figbirds, the raucous migratory koel and channel-billed cuckoos. Flying foxes squabble over the same food at night. Rainforest species such as the orange-footed scrubfowl and the gorgeous rainbow pitta also live here. When the region dries out, aggregations of butterflies seek refuge in these shady moist areas. The adults cluster in thousands, hanging from the leaves and branches of trees, waiting for the onset of the first storms. At least three or four different species can be seen, all members of the milkweed (danaid) family of butterflies.

OPPOSITE: A male saltwater crocodile in the shallows of the Yellow Water (Ngurrungurrudjba) Billabong.

FOLLOWING PAGES: Dusk at Ubirr, with the light of the setting sun reflecting off the Nadab floodplains.

I first visited Kakadu National Park in 1989. Among many highlights was a deeply emotional experience when I sat on a rocky platform with Elder Bill Neidjie on his own Country, at the edge of extensive wetlands, while he explained what it meant to him to be on Country. His passion for the place was palpable, as was his clear desire to share it. I understood the generosity of the Indigenous people in sharing such a magnificent place with the rest of the world. This struck me as a testament to the meaning of World Heritage.

Later, I watched a sunset from Ubirr and was awed by the timeline of the land, as I gazed across the plains that had for so long been the home of spectacular wildlife and the first Australians. I had goosebumps appreciating all that had gone before and a great hope that this place could continue to provide inspiration and meditation for future generations. Sadly, Bill Neidjie died in 2002.

OUTLIERS, KAKADU NATIONAL PARK PHOTO: ALLAN FOX

THE OUTLIERS

Once part of the Arnhem Land Plateau, the outliers were left behind as erosion progressively moved the escarpment eastwards.

1 When the explorer Leichhardt camped here in 1845, he "enjoyed the pleasing sight of large lagoons, surrounded with mangrove myrtles, with Pandanus, and with a belt of trees and Nelumbiums (lotus) ....."

2 The softer sandstone layers have been etched out by weathering.

3 The spiny spinifex provides shelter for small mammals and reptiles as well as food for termites.

4 Pandanus has an elaborate prop root system and is valued by Aboriginal people because the fruit is edible and the leaves can be stained and woven into baskets.

5 Freshwater mangroves fringe the edges of channels and billabongs and their leaves were used by Aboriginals to drug fish so that they could be easily caught.

KAKADU NATIONAL PARK

PRODUCED BY

AUSTRALIAN NATIONAL PARKS & WILDLIFE SERVICE

PATH TO WORLD HERITAGE

OUR KNOWLEDGE OF KAKADU'S natural and cultural heritage has continually expanded since its importance was first realised. Although the high natural quality of the Alligator rivers area of the Northern Territory had been formally identified as early as 1965, few Australians understood the significance of the area when the Northern Territory Reserves Board proposed the establishment of a national park. Justice Edward Woodward, as Aboriginal Land Rights Commissioner in the early 1970s, proposed a future that would enable Indigenous land rights and conservation to proceed together. It was his report that alerted the nation to the exceptional qualities of this area.

The trigger for action was a proposal to mine uranium deposits recently found in the area. The Ranger Uranium Environmental Inquiry (also known as the Fox Report) was set up to consider the possibility of uranium mining in the area and the likely effects on both the environment and the Indigenous people. At the same time, the Commonwealth *Aboriginal Land Rights (Northern Territory) Act 1976* envisaged the recognition of traditional ownership. This led to an agreement that granted land rights (excluding the Ranger Uranium Mining Lease) to the Traditional Owners, who in turn leased the land back to the Australian Government for the purpose of a national park.

OPPOSITE: A poster describes the 'outliers' of Kakadu—rocky areas that have been separated from the Arnhem Plateau by erosion.

ABOVE: The Movement Against Uranium Mining (MAUM) rally in 1979 protested uranium mining, particularly in the Northern Territory near Kakadu.

The implementation of these recommendations involved the declaration of Kakadu National Park in three stages. The first stage entailed establishing a board of management made up of a majority of Traditional Owners with a Traditional Owner as chair and a strong focus on conservation and tourism. The second and third stages involved the expansion of the park through acquisition of land. The final adjustment, at the longstanding request of the Traditional Owners, was the addition of the Koongarra mine lease site to the World Heritage area (accepted by UNESCO in 2011).

Kakadu National Park is managed by a board that must comprise an Indigenous chair and Indigenous majority. This ensures that the commitment to conservation of cultural and natural heritage is strongly observed in the development of policy. The area faces a number of threats, including the Ranger Uranium Mine, which stopped operating in 2021 but will continue to need work to remove toxins and rehabilitate the environment. Impacts from climate change of significant concern include progressive saltwater intrusion into lowland wetlands. Introduced species, such as invasive weeds, the cane toad and the feral cat, pose ongoing threats and impacts, and are being addressed by a focus on habitat management, including effective fire management. Despite these threats, the environment remains in good condition, ensuring that ecological and cultural processes thrive. Significant investment is needed to preserve the artwork and to ensure the sustainable presentation of the cultural and natural heritage attributes to the wider community.

As with most World Heritage areas, the board needs to manage large numbers of visitors, with a typical annual visitation of around 200,000 people. Dedicated walking trails minimise the damage to the area from foot traffic and pollution, as does limiting access to specific waterfalls and rock pools of the sandstone escarpment. Making only a select number of rock art galleries accessible limits the likelihood of degradation by vandalism. The wetlands can be safely enjoyed on boat trips and excellent visitor centres teach more about this magnificent place.

OPPOSITE: Controlled burning in Kakadu in 1998. Indigenous fire management strategies are used, along with western science, to benefit both biodiversity and traditional culture.

ABOVE: Murrumburr Elder Violet Lawson with lotus lilies threatened by salt contamination caused by rising sea levels.

GREAT BARRIER REEF

A pure white sandy cay, lying a long way offshore in the northern Great Barrier Reef, is surrounded by the emerald greens and azure blues of the ocean. One hot and humid summer, a tiny turtle emerged from the sand—so small she could fit in the palm of your hand. Baby flippers propelled her down the slope to the water's edge, where she tentatively slid into the ocean. She swam away quickly to clear the shallows, where danger lurked—herons awaited the annual feast on unlucky hatchlings. Further out, there were more risks: sharks were waiting, as were trevally, ready to take the hatchling whole. What happened to this beautiful but vulnerable green turtle? She made her way to the open ocean and out into the Coral Sea. Apart from finding food to grow, she also needed to avoid the many predators that threatened her survival. Over the next decade or so, she might have moved as far afield as the Solomon Islands and Vanuatu. Gradually, she grew and spent another three or four decades in the feeding areas of the Great Barrier Reef before returning to the beach of her birth to dig a nesting hole and lay her first clutch of eggs. Finally, a new generation would begin.

THE WORLD'S GREATEST BARRIER REEF extends 2,300 kilometres along the north-eastern coast of Australia, protecting the coastline from long-distance swells and storm waves and providing a mosaic of reef patches, islands and sheltered water. It also provides a barrier to shipping, as James Cook discovered in 1770 when the *Endeavour* ran aground on a reef south-east of present day Cooktown, in far northern Queensland. Within the Great Barrier Reef there are more than 3,000 individual reefs of varying sizes and shapes, providing homes for 600 species of soft and hard coral and 1,625 species of fish. There are also 3,000 varieties of mollusc and a great diversity of crustaceans, sponges, anemones and marine worms, as well as marine mammals. There is no other reef nearly as significant anywhere else in the world.

The marine park's 900 islands range from sandy coral and vegetated cays to rugged continental islands, such as the outstanding Hinchinbrook Island near Ingham, with its rainforests and mangroves, to beautiful tropical island groups, such as the Whitsunday Islands near Mackay. In the far north, Raine Island is the world's largest green turtle breeding area. The many islands north of Cardwell provide protected breeding sites for the pied imperial-pigeons that fly each day to the mainland to feed on rainforest fruits.

The Great Barrier Reef is the world's largest coral reef ecosystem, with examples of all stages of reef development. In places, one can step from a sandy beach backed by rainforest out to coral reefs within a few metres. Elsewhere, it may take a few hours sailing to reach the outer reefs. Extending over a full ten degrees of latitude, the Great Barrier Reef encompasses many different ecosystems, some strongly tropical and others more subtropical. It also extends across the continental shelf, from shallow inshore fringing reefs to outer reefs that in places drop off into water more than 2,000 metres deep. It is generally narrower in width in the north and wider in the south, varying between 60 and 250 kilometres across, from the shallow coastal low-water mark to the edge of the continental shelf. It captures an impressive array of mangrove species (over half the global species and 20 of the 21 families) that offer critical habitat for fish and other marine organisms. Throughout the area are many seagrass meadows, which provide habitat for the threatened herbivorous dugong. These strictly marine, long-lived and slow-breeding mammals are rarely found away from their food sources and may often be seen in coastal areas of the Great Barrier Reef.

There are two scales at which to best appreciate the splendour of the Great Barrier Reef. The first might be called the astronaut perspective—the view of this huge reef system from space. Such a view became possible only in the past 60 years and was a feature of early space exploration.

PREVIOUS PAGES: A turtle dives down to the reef in a shallow lagoon near Lady Elliot Island.

OPPOSITE: Aerial view of the Great Barrier Reef near Gladstone, Queensland, 1996.

ABOVE: A dugong feeding alongside golden trevally.

The area of the reef—344,400 square kilometres—is about the same size as 70 million football fields or, to put it another way, as Italy, Japan or Malaysia. This astonishing extent can only be fully appreciated from a great distance. With public access to satellite imagery, such views have become much more familiar but no less stunning.
The second scale is that of a diver encountering the reef close up. Although the coral reefs make up only seven per cent of the World Heritage area (the rest is open ocean, lagoons and islands), they provide a focus for the diversity of marine life found throughout the site and attract divers from all over the world. Every snorkel or dive reveals a sample of different corals, associated fish and other organisms that depend on the corals, and a cross-section of the remarkable structures that make up the reef.

This World Heritage area is home to many iconic wildlife species, but perhaps the most significant are the marine turtles and the cetaceans. Six of the total seven living species of marine turtle live in the Great Barrier Reef and most of these depend on the reef for food and breeding resources. The loggerhead turtle is critically dependent on the wide feeding areas as well as many breeding sites, such as Mon Repos near Bundaberg and several offshore islands. The Australian continental shelf is the only habitat of the vulnerable flatback turtle, which stays close to the shore, eschewing the open ocean migration of other species. The reef is one of the few places where the globally endangered hawksbill turtle is safe from human tortoiseshell hunters. The green turtle, while truly global in its distribution across subtropical and tropical oceans, finds the best protected breeding sites on this reef. Across the world, many thousands die as bycatch for fishing industries along with the tens of thousands harvested for human consumption. Green turtles are herbivores and, like the dugong, thrive on this area's healthy seagrass beds, which are better protected from human impact than any other green turtle habitat in the world.

Pixie Pinnacle on the Ribbon Reefs near Cooktown, Queensland.

One of my favourite sites in the Great Barrier Reef is the exquisite Pixie Pinnacle, on the Ribbon Reefs between Port Douglas and Cooktown. This small bommie (outcrop of coral reef) provides a spectacular dive, taken in a slow spiral from the base at around 25–30 metres drifting up to the narrow peak just below the waves. I like to go very slowly, concentrating on the schools of fish that accompany us, soaking up the gorgeous corals of many kinds, branching Acropora species, plate corals, large brain corals and stony forms, sea fans, numerous crinoids with waving arms and clumps of anemones with their beautiful clownfish. One goal is to find some of the colourful but tiny nudibranchs—soft-bodied hermaphrodite molluscs with exposed gills that come in a vast array of sizes, shapes and colours.

Drifting slowly up the face of the bommie reveals many different features, including an occasional stonefish, many butterfly fish and other small colourful fish, sometimes larger lionfish, moray eels or a big potato cod, small blennies and distinctive prawns and crabs hiding within the corals. Schools of fusiliers and other species circle off with watchful eyes looking out for reef sharks. On one occasion on a dive at the nearby Lighthouse Bommie, we were astonished to see three dwarf minke whales swim by, giving us the eye.

IN PLACES, THE REMNANTS OF INDIGENOUS STONE FISH TRAPS AND SHELL MIDDENS CAN BE SEEN, REFLECTING THE LONG CULTURAL TRADITION OF THE BANDJIN PEOPLE WHO LIVE ON THE ISLAND.

At least 30 species of cetacean (a group of aquatic mammals that includes whales and dolphins) occur in the Great Barrier Reef World Heritage Area. Humpback whales migrate from Antarctica to the reef each winter to calve and have been recently increasing in number as they recover from the population crash caused by late nineteenth- and twentieth-century whaling. The east Australian population now exceeds 10,000—a remarkable recovery from the 500 or so left at the end of mass whaling. Two resident dolphins, the Australian snubfin and the Indo-Pacific humpback, are a high priority for conservationists due to their vulnerability as inshore mammals (a habitat that receives the most impact from human activity). It has only been discovered recently that one small whale, the dwarf minke, uses the reef in winter months, probably for breeding, as small calves have been seen. This species, or at least some of the population, mostly adolescents, appears to display curiosity towards humans and many hundreds have now been recorded interacting with snorkellers and sometimes approaching boats. It is now known they also migrate south at the end of winter and travel into the Antarctic Ocean.

The islands and cays support an extraordinary diversity of birds, including critical breeding sites for many seabirds. Michaelmas Cay, some 40 kilometres off the coast of Cairns, provides an astonishing spectacle, with up to 20,000 breeding pairs of seabirds covering the two-hectare island. Sooty terns, crested and lesser crested terns and common noddies all nest on the sandy surface, feeding their young from the productive seas around the cay. Marauding frigatebirds, also in search of a meal, will frequently take advantage of the terns' hunting skills, attacking them to steal their catch. A number of other species are also known to visit, including boobies, other terns and some waders. As the birds wheel around the island in the thousands, calling to their mates or their young, they provide a superb visual phenomenon with accompanying rowdy sound. Two other coral cays with significant wildlife visitation can be found in the remote far south of the reef. Lady Musgrave Island (14 hectares) and Lady Elliot Island (45 hectares) are seabird-nesting islands surrounded by well-developed coral reef systems. Noddies and terns nest here each year, as do the migratory wedge-tailed shearwater and the uncommon red-tailed tropicbird.

In sharp contrast to these small, low-lying landmasses, the majestic and extensive Hinchinbrook Island rises to 1,100 metres above sea level. Several rugged mountain tops and slopes are covered with rainforest, patches of paperbark forest and heath, descending to extensive mangroves along the western and northern margins. This virtually untouched wilderness island matches that of the forests of the nearby mainland, providing spectacular vistas across the adjacent reefs and sea and back towards the mainland coast and ranges. The 50-kilometre-long Hinchinbrook Channel provides rich fish habitat and one of the most diverse mangrove communities in Queensland. In places, the remnants of Indigenous stone fish traps and shell middens can be seen, reflecting the long cultural tradition of the Bandjin people who live on the island. There are many other examples of historical and current Indigenous sites elsewhere in the World Heritage area. About 70 Traditional Owner groups have authority for sea country management within the World Heritage area from the Torres Strait island people of the north to the Gurang in the south.

OPPOSITE: Blue-green chromis fish often feed near coral.

FOLLOWING PAGES: The dwarf minke whale visits the Great Barrier Reef during winter.

The present form of the Great Barrier Reef is quite young, with its beginnings some 6,000–8,000 years ago when the sea reached present levels. Corals began growing and spreading when the sea level rose 20,000 years ago. Before then most of the Great Barrier Reef World Heritage Area was land and the reef was a narrow ribbon on the seaward edge of the broad plain.

It was only with the rising sea levels of the post-glacial era that the marine environment was established and corals were able to colonise the islands as fringing reefs and parts of the intervening areas where hard substrates survived. The history of evolution within the Great Barrier Reef is captured in coral structures that reveal past processes—including coral growth, climate and the effect of weather—and form part of the reason for the Great Barrier Reef's significance.

One of the most spectacular events in the reef's calendar is the annual, synchronised mass spawning of most species of coral. This mainly occurs on inshore reefs just after the full moon in October, while the outer reefs spawn in November or December (spawning is linked to sea temperature). The astonishing synchronisation was discovered about 30 years ago and overturned a long-held theory about how corals reproduced. Each coral produces both eggs and sperm, which are released at night, with different nights used by different species and entire species spawning in the same night. The scale of the mass spawning produces a cloudy effect as billions of eggs and sperm fill the water with different colours. The phenomenon only lasts a few days at each locality; the fertilised eggs form larvae, which then briefly drift in the currents before settling on the floor of the ocean.

OPPOSITE: Acropora coral spawning near Magnetic Island.

ABOVE: A close-up view shows individual eggs leaving the polyps of the staghorn coral.

PATH TO WORLD HERITAGE

THE GREAT BARRIER REEF moved rapidly from a holiday destination for a few people to a national conservation concern in the 1960s when the Queensland Government granted permits for oil exploration over large areas of the reef. Eventually, the permits gave access to almost the entire reef area and mining leases covered most of the inshore areas. Poet Judith Wright joined with conservationist John Busst to lobby the state and federal governments to overturn this attempt to drill for oil because of potential environmental impacts. Wright's book *The Coral Battleground* documents the many challenges that eventually led to a successful outcome and the establishment of the Great Barrier Reef Marine Park by the federal government in 1975. The *Great Barrier Reef Marine Park Act 1975* specifically prohibits mining anywhere in the marine park. Community concerns for the long-term protection were widespread and successive federal governments have maintained a commitment to the protection of the reef.

In 1981, the Great Barrier Reef became one of Australia's first three sites inscribed on the World Heritage List. At the time, the IUCN made the point that if there was to be only

OPPOSITE: Stand-up paddleboarding around Magnetic Island.

ABOVE: The crown-of-thorns starfish poses a significant threat to the coral of the reef unless removed.

one coral reef on the World Heritage List, it would have to be the Great Barrier Reef. Considerable support for the Great Barrier Reef also comes from the Australian Institute of Marine Science (established in 1972 near Townsville), which undertook significant research projects on marine biodiversity and which continues a monitoring program to check reef health. Similarly, James Cook University developed a globally significant marine biology program that continues to provide expert research and analysis on the reef systems. The Great Barrier Reef Marine Park Authority has legal responsibility for the World Heritage area and has developed innovative programs to monitor and report on the health of the marine ecosystems.

The 2014 Great Barrier Reef Marine Park Authority report revealed that the greatest threat to the health of the Great Barrier Reef comes from climate change. The IUCN reinforced this analysis in their 2017 Conservation Outlook report on the reef. UNESCO has considered listing the area as 'in danger' as a consequence of the current condition of the reef. Other threats—from sediment, chemical and nutrient run-off from the mainland and from adjacent coastal development—remain a concern and are being addressed by national and state programs. The crown-of-thorns starfish outbreaks appear to be related to the nutrient status of inshore waters. One catastrophic outcome of climate change has been the occurrence of coral bleaching in 2016 and 2017, and unprecedented coral bleaching in the El Niño year of 2020. There is a high risk that this will lead to significant mortality, from which many corals will not recover. In 2022, a UNESCO/IUCN monitoring mission to the Great Barrier Reef concluded the damage was already critical, especially water quality related to terrestrial run-off, and it made a recommendation for the site to be placed on the list of World Heritage in Danger. At the time of writing, the prognosis for the reef is grave but there are significant actions underway that could be effective in mitigating further damage.

OPPOSITE: Bleached coral in the waters off Great Keppel Island.

ABOVE: Biologist Dr Robert Endean examines a crown-of-thorns starfish.

WILLANDRA LAKES REGION

Standing near Lake Mungo today, you have to use your imagination to visualise how it may have looked 40,000 years ago: one of many deep freshwater lakes, overflowing with fish and shellfish, edged by vegetated sand dunes, with people living around its shores. The Paakantji, Ngiyampaa and Mutthi Mutthi people hunted small mammals and fished in the lake. They used stone tools, brought from far away, for hunting and food preparation. Already, people had lived here for thousands of years and practised a culture that included ceremonial burial. From first arrival, conditions had been stable, with biological abundance that meant people thrived here. They cooked meals on campfires, eating bandicoot, hare-wallaby, rat-kangaroo and several aquatic animals, including yabbies and mussels as well as perch and cod. People most likely knew the giant kangaroo; perhaps they also occasionally saw the huge herbivorous *Zygomaturus* and other large mammals, but it is likely that the stout flightless bird genyornis was already a creature of myth. This time of plenty would change. Lakes started to dry out and turn salty; rainfall declined and vegetation changed. Eventually, the climate cooled and became dry and windy. Woodlands became grasslands and dust storms prevailed. Perhaps some groups moved away, northward following the retreating waters. It became harder to survive, but many did, applying new skills to the changing environment. The record shows people continuously resided here for at least 42,000 years and possibly much longer.

SCIENTIFIC STUDIES REVEAL that the ancestors of the present Indigenous communities at Willandra first arrived at least 45,000 years ago, and possibly up to 60,000 years ago, part of the most astounding early human migration across the world, from Africa via India and south-east Asia. Those early Australians quickly dispersed across the entire continent and their descendants stayed mostly where they first settled. They thrived on the excellent conditions that prevailed at the time, but later descendants had to endure and adapt to an extended period of significant climate change. As the global conditions moved towards a glacial maximum, locally the lakes turned salty and dried out as the climate became much drier and cooler. The lakes never filled again and, although conditions have improved a little over the last 10,000 years, survival in this environment remains a serious challenge. A relatively stable, semi-arid environment has become the norm. The environment has preserved an outstanding record of both the natural changes over that time and the cultural responses to those changes. The area includes the world's earliest known cremation and ceremonial burial sites, afforded to Mungo Woman and Mungo Man 40,000 years ago. Evidence of developing stone technology and the importation of ochre and tools, the extensive use of freshwater resources and a long history of hunting small mammals is an outstanding testimony to human development through time. As long as 18,000 years ago, people here were collecting grass seeds and using grindstone technology to make flour.

Apart from the archaeological evidence of the people who lived here, the Willandra Lakes Region World Heritage Area also provides exceptional evidence about the way natural environmental processes have progressed and evolved over the last 100,000 years. Palaeontologists, biologists and geomorphologists are able to describe in considerable detail the process of change before and during the last great

PREVIOUS PAGES: The topography of the Willandra Lakes World Heritage Area includes dramatic rock formations.

LEFT: A rock formation at the Walls of China, Lake Mungo.

glacial period. The geological processes of deposition and erosion of sediments both preserve and expose details of the past. Climatic conditions have enhanced preservation and that has allowed palaeontologists to uncover the evidence of past ecosystems and human life. With changing climatic conditions, the site became more or less suitable for human occupation but the archaeological evidence shows that people have always lived here. Many outstanding scholars have studied at Willandra and their findings have resulted in significant progress in understanding both natural processes and human adaptation.

The once lush environment of freshwater lakes and surrounding woodlands is today a series of dry lake beds and surrounding dunes. The eroded lunette of Lake Mungo, called the Walls of China, developed over millennia and preserves an exceptional suite of archaeological and palaeontological evidence about the past. Many fossil records of animals have been preserved, including some of the now-extinct megafauna and many smaller mammals lost since European settlement, such as the eastern hare-wallaby.

There is desert beauty at Willandra now, with sand dunes, shaped and moved by wind, surrounded by bluebush, saltbush, chenopod and spinifex. The limited woodland includes eucalyptus mallee shrubs and several cypress species with a few wattle and hakea species. Each of the old lakes has an associated lunette—a geomorphic feature peculiar to Australia's semi-arid climate. A series of wet and dry periods over millennia produced these distinctive crescent-shaped features.

More recent discoveries at the World Heritage site include well-preserved human trackways of around 500 footprints dating back 20,000 years. Made in claypans, they survived because they were covered by windblown sand that preserved them; they are the oldest found in Australia and the largest collection anywhere in the world.

OPPOSITE: Saltbush, bluebush and spinifex are widespread at Willandra.

ABOVE: Human footprints at Willandra Lakes are the oldest found anywhere in Australia, dating to about 20,000 years ago.

FOLLOWING PAGES: Erosion formations at Lake Mungo.

My father's ancestors came out from England and Scotland in 1830 and I would often proudly tell our children that they were sixth-generation Australian on both sides. The feeling of being Australian was important and strong for me. As a professional geographer, I knew of Dr Jim Bowler's work at Willandra Lakes and I remember hearing and reading about the astonishing finds and the early revelations about Mungo Man. Suddenly, my focus on ancestry seemed wishful thinking at best and the longstanding claims of Indigenous people took on new meaning. Much more recently, DNA work has shown that the first humans to arrive in Australia quickly settled all the way across the continent and their descendants have been resident in much the same areas for over 40,000 years. That is about 2,000 generations! Every place studied has the same pattern. No wonder present-day Indigenous peoples say that they have lived on their Country since time began.

PATH TO WORLD HERITAGE

A YOUNG GEOMORPHOLOGIST, Jim Bowler, set out in the 1960s to study the Willandra area as an extension of his analysis of the landscape and climate in the Pleistocene epoch (between 1.8 million and 10,000 years ago). He identified from aerial photographs that Lake Mungo would be the best place to focus his attention: erosion on the lake's lunette had exposed the layers of sediment necessary for his study. His early excavations revealed shellfish remains and what might have been stone tools, while a return visit in 1968 uncovered what he thought might be burned bones. Subsequently, archaeologists John Mulvaney and Rhys Jones discovered a human jaw that was later confirmed to be the remains of a human female, now known as Mungo Woman. In 1974, Jim Bowler and Alan Thorne excavated the almost-complete skeleton of an adult male—Mungo Man, who was buried with ritual use of ochre.

Dating to 42,000 years before the present, these are some of the oldest human remains in Australia and some of the oldest modern humans in the world outside Africa. Not only did the deposits at Willandra allow a comprehensive interpretation of evolving landscape processes, but they gave a significant insight into the lives of human communities over an enormous time span. It was on this basis that the Australian Heritage Commission in 1981 prepared a World Heritage nomination for a mixed site—both natural and cultural heritage. New material continues to be uncovered, including 20,000-year-old footprints, which have added value to the site. It is now known that Indigenous people have lived continuously at Lake Mungo for at least 45,000 years—a time span that represents around 2,000 generations. Both the extended history of occupation and the dry environment have combined to create outstanding archaeological heritage. The unanticipated discovery of the great antiquity of humans in Australia overturned existing ideas about Aboriginal prehistory and lent support to the claims of Traditional Custodians. Mungo Woman and Mungo Man caused a very significant revision of the science and together brought us a clearer vision of the past. This in turn triggered a World Heritage listing.

Paakantji, Mutthi Mutthi and Ngiyampaa peoples welcome visitors and share their stories.

OPPOSITE: A red kangaroo grazing on saltbush.

ABOVE: Elders of the Paakantji, Mutthi Mutthi and Ngiyampaa communities, the Traditional Owners of the Willandra Lakes area, examine Pleistocene human footprints at Lake Mungo.

LORD HOWE ISLAND GROUP

Two thousand metres below the surface of the South Pacific, a fury of molten lava burst through the mantle and vented upwards. Boiling rocks from the heart of the earth rose through the incredible pressure of the deep. Over time, there built an underwater mountain that eventually, perhaps seven million years ago, reached the air above the sea with much steam and forceful spewing of liquid rock. The mountain grew to a great height over a large area—a giant shield volcano. Slowly, ocean forces eroded the massive area until it was a tiny fraction of its original size: an island outcrop from an underwater plateau. Over time, the few plants and animals that found their way across the open ocean to the island evolved to become the biological treasures of today. Corals grow in the shallow waters around the islands—the southernmost coral reef in the world, with a mix of temperate and subtropical species. One of the most beautiful islands in the world, thought to be unknown to humans until 1788, it should disappear beneath the waves 200,000 years from now.

THE LORD HOWE ISLAND GROUP World Heritage Area is an outstanding example of oceanic islands of volcanic origin. The islands provide important breeding grounds for colonies of seabirds as well as critical habitat for threatened species of plants and animals. Although the terrestrial area is small (1,540 hectares), the mountains rise abruptly to 875 metres (Mount Gower) before plunging steeply into the sea. They impose a grandeur on the landscape that is in sharp contrast to the central area with its beaches and shallow inshore waters. The main island is dominated by palms and rainforest and almost half of the island's 241 species of plants are endemic.

Located some 700 kilometres north-east of Sydney and forming part of the territory of New South Wales, Lord Howe Island supports 370 residents and is managed by the Lord Howe Island Board. The first recorded sighting was in February 1788 by Captain Henry Lidgbird Ball while he was sailing the brig HMS *Supply* to Norfolk Island to establish a penal colony. Subsequently, whalers and others visited Lord Howe Island and some settled there, surviving on introduced goats and pigs, which ran wild. Within a short time, many of the endemic species became extinct, most from the effects of humans and the rats they introduced. It took a substantial period of time—a century or more—for people to appreciate the unique nature of Lord Howe Island, in terms of both its geology and biology, and to try to halt the extinction of its flora and fauna.

Many distinct vegetation communities give the island its rich biodiversity, from the mountain tops to the beaches. One of the most important communities is the unique palm forests, with four endemic species. Seeds from kentia palms formed an early part of European trade in the island's wildlife. These seeds were particularly valued because people wanted to grow palms elsewhere, especially

PREVIOUS PAGES: Looking south at Mount Lidgbird (far left) and Mount Gower.

OPPOSITE: Banyan trees and kentia palms.

ABOVE: Aerial view of Lord Howe Island and surrounding marine park, containing the southernmost coral reef in the world.

in Europe. Ferns are also abundant; 19 endemic species grow here. Temperate rainforests are prominent, occupying habitat associated with the two main peaks, Mount Gower and Mount Lidgbird. Near the summits, gnarled mossy cloud forests capture moisture from the clouds that cloak them for most of the year. Plants in this community, including ferns, and other rare and threatened species like the little mountain palm, depend on the continuing supply of moisture from the mists. This feature is extremely rare in small oceanic islands, which do not have the elevation that is usually necessary to capture or produce cloud moisture. The lower elevations of the island have subtropical rainforest, including the critically endangered Sallywood Swamp forest.

The high peaks of Lord Howe are important for other reasons. The astonishing giant phasmid—the world's largest stick insect—thrived in this area. A 15-centimetre-long insect sometimes called a land lobster, it was once prolific across the entire island but was almost rendered extinct by rat predation. However, 15 years ago a tiny population was found on the rat-free Balls Pyramid and there is now a captive breeding population on the Australian mainland, as well as on Lord Howe Island. The intention is to reintroduce them to Lord Howe Island once the rats have been eradicated. There is hope that such programs can help maintain at least some of the distinctive biodiversity once present on the island.

The beauty of the island is complemented by the large number of wildlife species found nowhere else in the world. Huge flocks of seabirds use the island for feeding, roosting and breeding. It is the breeding headquarters for the vulnerable providence petrel—around 30,000 pairs breed on the island. From a population of millions, the petrel was hunted to extinction on Norfolk Island in just three years following settlement. The species is now most abundant on Lord Howe Island. The world's largest breeding aggregation of red-tailed tropicbirds also nest here. They can easily

THE BEAUTY OF THE ISLAND IS COMPLEMENTED BY THE LARGE NUMBER OF WILDLIFE SPECIES FOUND NOWHERE ELSE IN THE WORLD. HUGE FLOCKS OF SEABIRDS USE THE ISLAND FOR FEEDING, ROOSTING AND BREEDING.

be seen flying off cliffs around the southern and northern mountains, displaying their long red tail streamers and red bill with sharply contrasting white body plumage.

The remarkable story of Lord Howe Island's land birds is one of survival and extinction. In common with many remote island ecosystems, bird ancestors arrived rarely and their offspring frequently diverged to become new forms and, in some cases, new species. When humans arrived, there were 15 species of resident breeding land bird, six of which were endemic species and eight of which were endemic subspecies. Nine of these unique species are now extinct due to the introduction of pests (rats in particular)

OPPOSITE: Observatory Rock (left), Wheatsheaf Islet (centre) and Balls Pyramid (right).

ABOVE: A providence petrel of Lord Howe Island.

W. T. Cooper. 71.

and the deliberate actions of residents (the endemic Lord Howe red-fronted parakeet was exterminated as an agricultural pest). The impressive white gallinule, a large flightless bird only found on Lord Howe Island, was eaten to extinction by people. Surprisingly, another flightless bird, the Lord Howe woodhen, survived, but only just. It was abundant and tame when the *Supply* first sent men ashore. Over the following 150 years, the population plummeted due to consumption by humans, predation by introduced cats and egg loss to rats, and habitat damage by pigs. By 1980, there were only 20 woodhens left. There has now been a remarkable recovery thanks to the efforts of the Lord Howe Island Board, NSW Parks and Wildlife staff and numerous volunteers, who have implemented cat, goat and pig eradication programs and captive breeding for the Lord Howe woodhen. Today, there are over 1,100 Lord Howe woodhens on the island, a wonderful success story. At the time of nomination this was thought to be the rarest bird in the world.

One of the first animals seen by humans on Lord Howe Island were marine turtles, most likely green turtles, which remain common in the waters around the islands. Some were collected for food when sailors from the *Supply* first landed on the island. They are not known to breed on the beaches there, but form part of the Queensland breeding population. Perhaps the most amazing turtle recorded on Lord Howe Island is the horned tortoise, a species which is long extinct but may have lived on the island as recently as 100,000 years ago. It was a land animal, with a horned head and a tail club, that grew to more than two metres in length and must have been an impressive beast as it roamed across the island. Its presence on Lord Howe Island (and that of its relatives on other oceanic islands) raises questions about the geological formation of the islands. How might such terrestrial animals have found their way to these mid-oceanic islands? Some of their relatives survived in Vanuatu up until the time of contact with humans. Apart from turtles, other large animals found around the island include marine mammals such as the bottlenose dolphin and humpback whale. The only terrestrial mammal is a small insectivorous bat.

An extensive marine park (state and Commonwealth) surrounds Lord Howe Island, containing the southernmost coral reefs in the world. A great variety of underwater topography is evident in the different ecosystems around the islands—from fringing reef to extreme drop-off features and the world's largest sea stack (Balls Pyramid). The corals are diverse, extremely colourful and include many tropical species at the southern limit of their range. Over 90 species are found here, along with 500 species of fish. Of special significance at Lord Howe is the unusual mixture of tropical and temperate species and the rarely seen transition between coral and algal reefs. As might be expected, there are many endemic species of marine invertebrates in this isolated ecosystem. Although studies are incomplete, there are 65 species of echinoderms known in the marine waters, including sea stars and starfish, sea urchins, sea cucumbers and feather stars. There are several endemic fish species as well, including McCulloch's anemonefish, the doubleheader wrasse and the Lord Howe pipefish.

Lord Howe Island is one of the world's most beautiful islands, thanks partly to the geological processes that created the landforms and the relatively undisturbed environment (about 70 per cent remains largely untouched). The unique assemblage of terrestrial and marine wildlife has added further to its character. This combination of beauty and biodiversity is at the heart of Lord Howe's outstanding universal value and World Heritage listing.

OPPOSITE: Orange-fronted, red-fronted and yellow-fronted parakeets. The Lord Howe red-fronted parakeet was deliberately exterminated by early settlers, who considered it a pest.

FOLLOWING PAGES: Blackburn (Rabbit) Island sits amid the reef in the lagoon.

When I was a student at the University of Western Australia, I read a wonderful book on the theory of island biogeography, written by Robert H. MacArthur and Edward O. Wilson. This totally captured my imagination and led to my many interests in ecology and wildlife. The book set out to explain why species richness differs between islands and why so many island species are endemic. Were there rules or laws that could be used to predict the occurrence of species diversity on islands? I have since spent a lot of time on islands all around the world exploring their diversity.

Islands provide excellent insight into the management of protected areas, which are themselves sometimes characterised as 'islands in a sea of agriculture' or similar. My focus was often on why so many island species become extinct. Extinction rates are very high on islands, especially following human contact. Of all the birds that have become extinct in the world over the past 200 years, over 90 per cent have been on islands. Lord Howe Island has lost several since humans arrived—an all too familiar story for those who love islands. It is therefore uplifting to consider the Lord Howe woodhen and its return from near extinction.

PATH TO WORLD HERITAGE

A Lord Howe phasmid from the island's captive breeding population.

THE IMPORTANCE OF the Lord Howe Island environment was recognised quite late in the history of settlement and it was not until 1981 that the New South Wales Government established it as a permanent park preserve. This preserve covered only terrestrial areas and the protection of their flora and fauna. In 1982, the Australian Government successfully nominated both aquatic and terrestrial areas for World Heritage listing based on their astonishing beauty and the level of diversity and endemism among the areas' species. In 1999, a marine national park was declared under New South Wales law, which increased protection of the marine environment. A Commonwealth Marine Reserve was declared in 2002, which covers most of the World Heritage area. All of the World Heritage area is protected under the *Environment Protection and Biodiversity Conservation Act 1999*.

The Lord Howe Island Board is responsible for the day-to-day management of the islands and it has played an important role in working for the conservation of the remarkable features of the World Heritage area. The board has sponsored major efforts to save the woodhen as well as earlier ecological studies that formed the basis of the recovery plan. It continues its far-sighted work with investment in weed eradication and a volunteer program. Since 2004, over 150 volunteers have contributed 16,000 hours towards weed eradication. In 2019, Lord Howe Island was the largest populated island to undertake full-scale rodent eradication. Extensive post-eradication monitoring and community reporting identified a small population of rats in 2021, which was removed and the two-year post-eradication monitoring period was restarted. Biosecurity has substantially increased following the eradication and continues to improve with greater stakeholder and community engagement. Many biodiversity benefits have been observed since eradication and a monitoring program is underway to document these.

The Lord Howe woodhen has been returned from the brink of extinction through a captive breeding program.

TASMANIAN WILDERNESS

Twenty thousand years ago, an orange-bellied parrot fed along the coast of South Australia, not far from what is now the Victorian border; a relatively cool spring was ending. Gradually, the bird flew east and followed the coast south, stopping periodically to feed on the extensive button grass meadows. Its journey took it well south, down the west coast of Tasmania into an area where tough eucalypts survived in a coastal strip, removed from the inland glaciers at higher elevations. Here, it sought out hollow limbs in which to build its nest, while enjoying the moorland food resources. Summer conditions were warm, and the parrot had three months to raise its young before the cold set in and it returned north. We can only imagine the challenges to this bird's descendants over the next 12,000 years as the land bridge slowly disappeared beneath Bass Strait, forcing each generation to fly longer distances across water during their annual migration.

UNESCO USES TEN CRITERIA to determine whether a place is of sufficient significance for inscription on the World Heritage List. A property only needs to meet one of these criteria to be accepted. Some properties meet several. The Tasmanian Wilderness World Heritage Area meets seven criteria, the most of any World Heritage area in Australia. Only one other site out of over 1,000 in the world has also met seven criteria: Mount Taishan in China.

The rugged geological landscape of the Tasmanian wilderness displays the effects of severe glaciation over many periods in the last 34 million years. It is an exceptional southern hemisphere record of glacial processes, features and landscape evolution—a significant reason for this magnificent area's inscription on the World Heritage List. During the most recent glacial maximum, ice covered much of the highlands of Tasmania and, in the course of formation, movement and retreat, has shaped much of what is present today. Glaciers abraded the landscape, scouring out valleys and forming lakes as they flowed slowly from the mountains and again as they retreated in a warming climate. Dove Lake is an excellent example of a glacial lake. One of the most appealing scenes in this landscape is the view across Dove Lake to the rugged Cradle Mountain skyline behind. Here, there are snow gums, pencil pines and deciduous beeches. Wombats, pademelons and the thickly furred Tasmanian short-beaked echidna live among the tussock grasses. The larger Lake St Clair was scoured by glacial processes to become Australia's deepest freshwater lake (167 metres). The 65-kilometre, six-day hike along the Overland Track from Cradle Mountain to Lake St Clair is one of Australia's most famous walks, traversing a spectacular, rugged, glacier-shaped wilderness. Further south, Mount Field was also covered by a snowfield that fed several glaciers in the surrounding valleys, producing the distinctive landscape features of today. The mountain ranges and peaks in this World Heritage area are dotted with tarns and dissected by wild rivers that mostly retain their pristine conditions, a rare phenomenon.

The exceptional natural beauty of the mountains is highlighted by wild and rugged coasts; stretches of sandy beaches; button grass, heath, alpine and subalpine communities; and the largest area of temperate rainforest remaining in Australia. Stately ancient Huon pines stand tall in the sheltered lower reaches of rivers and, elsewhere, towering eucalypts make up the world's tallest flowering forests. Individual trees are majestic; collectively they are awe-inspiring. There are significant links between the endemic plants and animals of the area and the former Gondwana. The Darwin Crater site contains an extensive pollen record of Pleistocene flora, from which scientists have been able to glean new understanding about the nature and evolution of vegetation communities.

The vast size of the property covers 1.58 million hectares, representing 23 per cent of the land area of Tasmania. There are few other places on the planet with such a large area of undisturbed forest habitat. Within Australia, this World Heritage site is synonymous with wilderness and the associated human experiences. The scenic beauty of such an extensive area is dependent on its unspoilt nature. Unsettled and undeveloped coastlines are rare anywhere, but these are particularly spectacular because of the surrounding hinterland that retains its ancient characteristics. The rivers flow clean and wild from the high-rainfall mountains through remote, undisturbed valleys to ultimately discharge into wild coasts and seas. It is not just the challenge of an untamed environment that attracts people, but the chance of wild serenity, remote from the noise, pollution and bustle of urban life. It is possible to hike for days through this magnificent wilderness and feel at one with natural processes. Few other places provide their visitors with such inspiration and opportunity for reflection.

The Tasmanian wilderness area protects a variety of endemic plants, animals and intact interconnected ecosystems, and provides significant natural habitat for the conservation of biological diversity.

PREVIOUS PAGES: Lake Oberon, Western Arthur Range, Southwest National Park.

OPPOSITE: Dove Lake, looking to Cradle Mountain, Cradle Mountain–Lake St Clair National Park.

Many of these species are of great importance for conservation and science, such as the critically endangered orange-bellied parrot, mentioned above, and the unique Tasmanian devil. The eastern quoll is now extinct on the Australian mainland but still found in this World Heritage area, as is the spotted-tail quoll. Until last century, the thylacine (Tasmanian tiger) roamed this area; today we lament its loss. Wombats remain in good numbers and the Tasmanian pademelon is also widespread. Most of the endemic species of bird are also found in the World Heritage area, including the dusky robin, yellow-throated honeyeater, scrubtit and yellow wattlebird. Tasmania is a world hotspot for freshwater crayfish; the unusual endemic burrowing freshwater crayfish has been linked to Gondwana fossils.

The most extensive and pristine areas of alpine vegetation in Australia occur here. Shrubs are the dominant plant species and some 60 per cent of the flora is endemic to Tasmania, including fragile cushion plants (six species in five families), the beautiful pink-flowered mountain shrub scoparia and the deciduous beech that gives superb autumn colours across the alpine areas. On mountain tops, snow gums grow only in sheltered areas; elsewhere, they form open woodlands, showing off beautiful coloured bark. The shrubby varnished gum, the world's smallest eucalypt, grows here. King Billy pines and other conifers add another dimension to the vegetation, especially in the montane dwarf rainforests associated with myrtles, leatherwood and the strange pandani—the largest heath plant in the world. A few rare sphagnum peatland areas grow on hillsides above 600 metres, forming a distinct ecological community in moist, poorly drained infertile soils. Some vegetation communities are fire-adpated and recover well following fires, such as button grass moorlands and eucalypt woodlands. Alpine vegetation and rainforest communities contain fire-sensitive species, such as pencil pines, Huon pines, King Billy pines, myrtle-beech, deciduous beech and sphagnum, that may take hundreds or even thousands of years to recover from a fierce burn.

SHRUBS ARE THE DOMINANT PLANT SPECIES AND SOME 60 PER CENT OF THE FLORA IS ENDEMIC TO TASMANIA, INCLUDING THE BEAUTIFUL PINK-FLOWERED MOUNTAIN SHRUB SCOPARIA.

The cool temperate rainforest is a highly valued element of this World Heritage area. Myrtle beech is usually the most dominant tree species. Its links to Gondwana explain why it once grew extensively on mainland Australia. It can grow to 50 metres tall and may live 500 years. Other tree species in these temperate rainforests include sassafras and leatherwood. Horizontal scrub is another tree species that forms a fascinating forest. As the individual trees grow, they bend over and then new shoots form numerous vertical branches that themselves bend over and repeat the process. This creates impenetrable scrub, much to the chagrin of bushwalkers. There are also many species of understorey plants, including climbing heath, with its beautiful pink bell-shaped flowers, which also has a Gondwanan link. Native laurels, restricted to high rainfall areas, forms understorey shrubs and stands out because of its large leaves, rare

OPPOSITE: *Richea scoparia* at Lake Salome, Walls of Jerusalem National Park.

ABOVE: A lone thylacine in the Beaumaris Zoo, Hobart, around 1930. The last known thylacine died in this zoo in 1936.

In February 2016, I spent two days in the remote but magnificent Melaleuca area in the far south-west. Melaleuca is mostly undisturbed heathlands and button grass plains and supports eastern ground parrots, beautiful firetail finches, southern emu-wrens and other species of wildlife. I had come with friends to find and photograph the orange-bellied parrot as well as enjoy the remote environment. We camped in the walkers' hut and spent the day searching for wildlife. My first sighting of the parrot was of two perched on a dead shrub. To me, they were exquisite and I trembled with excitement to see them. It was with some sadness that I contemplated their parlous situation. There is an encouraging increase in the number of returning birds recently, including more wild-hatched individuals. They survive on feeding stations provided in Melaleuca and captive breeding efforts elsewhere. Better protection of the migratory habitat outside the World Heritage area is essential if the species is to persist.

in Tasmanian rainforests. The rainforests also support numerous fern species that, at times, may dominate patches of forest and include epiphytic species (species that grow on the surface of another plant). Tall and beautiful tree ferns also grow here, giving a distinctive enclosed feeling to the forest.

Tasmania is famous for its giant trees, the tallest of which is mountain ash; it can grow as tall as 100 metres. The most recent expansion to this World Heritage area (2013) included the Styx Valley area, which is well known for its forests of these extraordinary trees. The tallest flowering trees in the world, they rival the giant redwoods of California but grow to their size much more quickly despite being hardwoods, which tend to grow slowly. There are now over 100 giant trees registered in Tasmania, mostly named, but not all are within the World Heritage area. It is not just the height of a tree that makes it spectacular, but also its girth and overall volume. Globally exceptional ancient trees also occur, including individual Huon pines more than 2,000 years old. These additions contribute substantially to the wealth of attributes of this World Heritage area.

Port Davey Marine Reserve, now part of the World Heritage area, is one of the world's most fascinating estuarine systems. Several rivers and creeks drain rust-coloured water into the reserve, creating a thick, dark-russet-coloured layer of tannin-rich fresh water on top of the tidal salt water. This inhibits the penetration of sunlight and limits the growth of marine plants. Instead, colourful marine invertebrates—beautiful and unusual organisms—are concealed below its stained surface waters, including deep-sea species present at unusually shallow depths.

One of many examples of remarkable life in this World Heritage area is king's lomatia. This extremely rare plant can only reproduce vegetatively, a weirdly counterintuitive condition for a flowering plant, which would usually reproduce by its seed. Although it looks like a holly, and is sometimes called king's holly, it is in the same family as the banksia, Proteaceae, and despite producing flowers

LEFT: Eucalypt trunk and ferns.

INSET: A captive breeding program supplements the meagre wild population of orange-bellied parrots at Melaleuca.

it cannot produce seeds. It is naturally sterile, a rare condition in nature. But fossils from nearby Melaleuca show that it was alive some 43,000 years ago, making the surviving clumps perhaps the oldest plants in the world and indicating persistence through the last ice age.

Many areas have extensive and rich cave networks that include, most significantly, records of human presence. These give evidence of the hunting and gathering activities of the people who were at the southern limit for human life through the last glacial epoch. The Kutikina Cave provided sufficient archaeological material for the recognition of an exceptional record of human life at the time of World Heritage inscription. Subsequently, other sites have extended the time of occupancy to 35,000 years ago. Some of the caves contain troves of well-preserved layers of animal bones, tools and hearth, while others show the remains of art, including ochre and hand stencils, as found deep in Kutikina. The stratified cave deposits preserve a record of human occupation and adaptation in the face of environmental change over long periods. These sites, and significant coastal middens, form the basis of Indigenous cultural heritage recognition by UNESCO and give an excellent source for interpretation of life over thousands of years.

Forty thousand years ago, when Indigenous people arrived in Tasmania, there was a land bridge—likely a broad grassy plain—connecting it to the mainland. Over the following 20,000 years, as glacial conditions continued to develop, the central mountains became increasingly dry and cold. Glaciers occupied the higher parts of the landscape and slowly ground their way down the stream and valley lines. The vegetation was mainly grassland rather than forest and palawa survived well on both the grass seeds and mammals grazing on the grass. They hunted the red-necked wallaby (Bennett's wallaby), still prevalent throughout the area. While they also hunted other animals, wallaby bones dominate those found at archaeological sites. It appears that Indigenous people may have used fire to maintain the valuable grasslands on which the wallabies fed. More recently, with post-glacial warming and increased rainfall, forests and moorlands replaced the grasses, which would have reduced the supply of wallabies in these upland areas.

To survive in these areas, in sight and sound of glaciers, the people not only sought refuge in caves but also made warm clothing from animal skins. Sharp-edged stone tools were used for cutting and skinning, some brought from far away, indicating the social connectivity within these areas. People fashioned bone into points so they could make holes along the edges of skins, binding them together with sinews to produce garments. Following increasing rainfall, declining grasslands and the isolation of Tasmania from the mainland due to rising sea levels, Indigenous people became much more focused on coastal resources. This coincided with the expansion of forests from their glacial refuges to occupy most of the mountain slopes. Middens along the west coast provide further examples of Indigenous culture. Some coastal communities created beautiful shell necklaces, the oldest of which dates to 2,000 years ago. Examples of these can be seen on the Needwonnee Walk at Melaleuca.

OPPOSITE: Davey River, Port Davey, south-west Tasmania.

ABOVE: A maireener shell necklace traditionally made by women.

PATH TO WORLD HERITAGE

STUNNING SCENERY AND UNIQUE WILDLIFE meant that, as in other Australian states, a number of natural areas in Tasmania were set aside as reserves. These included some core areas of what would later become the Tasmanian Wilderness World Heritage Area. Mount Field was declared as one of Tasmania's first two national parks in 1916, and soon after the spectacular Cradle Mountain and Lake St Clair areas in the north were protected. There was further progress in conserving these areas when Tasmania's National Parks and Wildlife Service was established in 1971 and after communities protested the construction of hydro-electric dams in the south-west.

One cannot tell the Tasmanian wilderness story without explicit reference to Lake Pedder and its tragic drowning by a hydro-electric dam. This exquisite natural feature is today preserved in the form of brilliant photographs that inspired a generation of Tasmanians to take action. The Lake Pedder Action Committee fought against the Tasmanian Government's decision to dam the Huon River and thereby flood Lake Pedder National Park. The Tasmanian Government went ahead despite the lack of popular support, and considerable national controversy. Many well-known figures were involved, including Bob Brown,

OPPOSITE: A red-necked wallaby near Kitchen Hut, Cradle Mountain–Lake St Clair National Park.

ABOVE: The welcome party for arrivals at the upper camp on the Gordon River, 1983, from which protesters did their best to prevent work on the Franklin River dam.

the artist–photographer Olegas Truchanas, Peter Dombrovskis, Max Angus, Clive Sansom, Brenda Hean, Dick Jones, Kevin Kiernan and many others. Conservationists continued to battle the Hydro-Electric Commission and the Tasmanian Government's plans for more dams in what was then the untouched south-west wilderness. The South West Tasmania Action Committee formed to spearhead the battle against further dams on the Gordon and Franklin rivers that would transform the wilderness.

The famous slogan 'No Dams' on a yellow triangular background soon gained national recognition and the Tasmanian Wilderness Society was formed in 1976.

ABOVE: Peter Dombrovskis' haunting photograph of Rock Island Bend on the Franklin River was used in Wilderness Society advertisements ahead of the 1983 federal election.

INSET: This 'No Dams' badge was created for the Franklin River campaign of 1983.

OPPOSITE: Franklin Range from Cleft Peak, Southwest National Park.

The Franklin River campaign was successful in raising support for protecting the wild river and it was on the back of that success, and the associated national campaign, that the name was changed to become simply The Wilderness Society. The legal basis of the federal government's success in the High Court (*Commonwealth v Tasmania* (1983)) was the external power granted to it under the Australian Constitution and the fact that Australia had ratified the World Heritage Convention. Subsequently, another challenge, this time from Queensland, reinforced the ability of the federal government to make laws in respect of World Heritage, which is now considered a matter of national environmental significance.

National recognition of the importance of the Tasmanian wilderness led to the initial world heritage nomination in 1982. The World Heritage Committee, while inscribing the site, expressed concern at the possible damage from continuing proposals for dams and drew attention to the impact of mining as well as other activities outside the area. Subsequently, the federal government submitted for listing a significant expansion that was accepted by UNESCO and included areas of old-growth forest on the eastern margins. There had also been a great increase in knowledge about the archaeological sites that further increased the cultural significance of the site.

Since the 1989 expansion, there have been continuing modifications to the area, as Australia has proposed including more old-growth forests, decided to remove mining from the World Heritage area (in 2010) and further expanded the area (in 2012 and 2013). Additional controversy arose in 2014 when the federal government moved to remove some of the forests from the World Heritage area to enable logging. In November 2015, UNESCO sent a reactive monitoring mission to examine the state of conservation in the World Heritage area. This was partly in response to expressions of concern about plans to build a dam and permit commercial logging on the property.

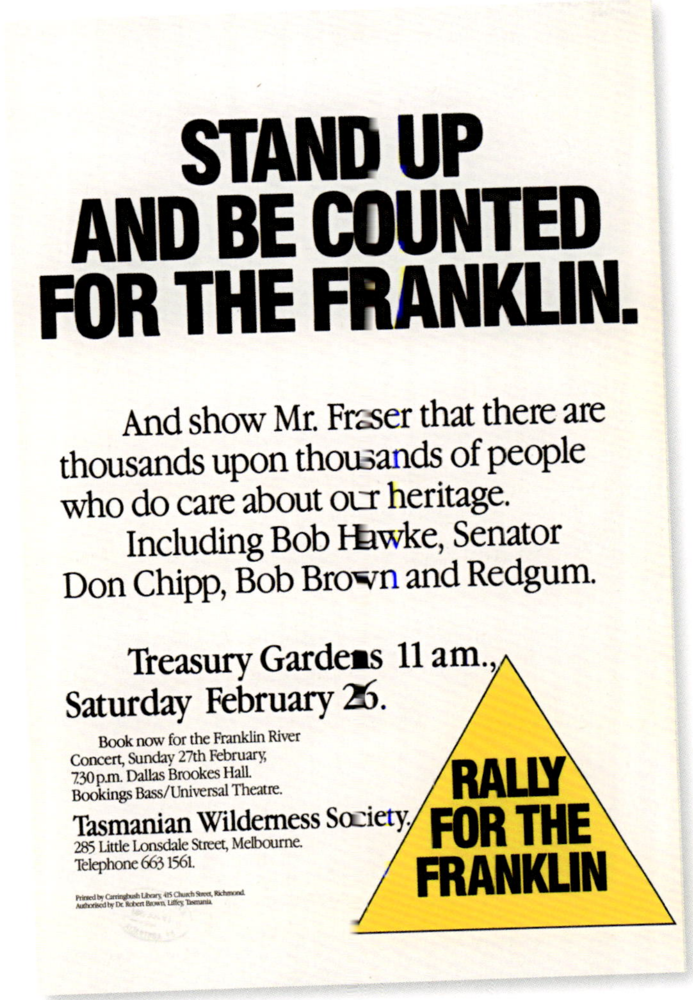

Other areas of concern were adjustments to management plans, mineral exploration and biosecurity. These matters have now been resolved to UNESCO's satisfaction, although there remain further community proposals for additions to the World Heritage area to protect these irreplaceable areas, notably in the Tarkine forest reserve.

A key threat for ongoing consideration is the impact of climate change, which has prompted the development of a comprehensive fire management plan to address heightened risks. Many also recognise the need for more detailed documentation of the area's cultural heritage, work that should occur in the near future.

PREVIOUS PAGES: Reflections, mists and *Melaleuca* trees in a serene Lake Pedder, 1968.

OPPOSITE: Dunes and granite near Interview River in the Tarkine wilderness.

ABOVE: Rallies protesting the Franklin dam took place around the country in early 1983.

GONDWANA RAINFORESTS OF AUSTRALIA

It is a serious challenge to visualise the continental scale transformations that created the world. The landforms of central eastern Australia seem so stable, so permanent, that the mountains, valleys, rivers and coasts must surely have been there forever. Seeing this region some 25 million years ago would be remarkable. Already, millions of years of erosion—after the continent separated from Gondwana—would have created the Great Escarpment which would have been retreating westward from the now sandy coastline. Over a period of perhaps three million years, numerous small, and some huge, volcanoes formed. Some created mountains, their slow-moving lava forming giant shields. Others burst violently forth in great explosions between long periods of quiet. Over the 20 million years to the present, these volcanic landforms gradually eroded, leaving behind great calderas in some places, and elsewhere shaping other distinctive features that rose above valley floors. Rainforests colonised the basalt rocks, soils formed after volcanism subsided and vegetation began to occupy the various niches. As the continent drifted further north and dried out, the rainforests retreated. The once widespread forests are now rare—a major transformation has occurred.

LOCATED ALONG THE CENTRAL EAST COAST, the Gondwana Rainforests of Australia World Heritage Area consists of 41 separate areas forming a series of rainforest remnants that extend from Barrington Tops National Park in New South Wales to Main Range National Park in Queensland. Ten of the patches are quite large (10,000 hectares or more), including well-known national parks such as Border Ranges, Washpool, New England and Lamington. Another seven are between 2,000 and 5,000 hectares in size, including Springbrook and Main Range national parks in Queensland and Wollumbin and Nightcap national parks in New South Wales. These separate elements cover a distance of 500 kilometres south to north; scattered over that area are another 24 smaller patches of forest. Together, these reserves capture the outstanding geological history of volcanism in this region, as well as the distinctive temperate

and subtropical rainforests that are remnants of the former Gondwana vegetation. Both geology and biology feature strongly in this World Heritage area.

When Australia separated from Antarctica, as the ancient continent of Gondwana broke up, new continental margins developed along the east coast. The associated swell (uprising) that ran parallel to the coast was subsequently eroded over many millions of years and formed both the Great Divide, mainly inland, and the Great Escarpment, closer to the coast. More recently (just over 20 million years ago), a series of volcanoes erupted sequentially and formed major shield volcanoes beginning with Tweed in the north and ending with Barrington in the south. The volcanoes were formed by the slow northward movement of the Australian continent over a hot spot in the earth's mantle. The Tweed shield volcano, centred on what is now Mount

PREVIOUS PAGES: The view from Thunderbolt Lookout in Barrington Tops National Park, New South Wales.

ABOVE: Nugurun Falls in Lamington National Park, Queensland.

Warning, subsequently eroded, leaving the best-preserved erosion caldera in the world. The central mountain mass (the former volcano plug) forms Mount Warning itself, which is surrounded by the eroded landforms, including the enormous Tweed Valley—what is left of the caldera. This impressive geological feature has enabled considerable understanding of these large-scale earth processes.

The volcanic lava gradually turned into basaltic soils that today provide fertile environments for the rainforests that occupy the reserves. These vegetation communities contain outstanding examples of the major stages of the earth's history. The Carboniferous period (around 300 million years ago) was dominated by ferns and their allies—this was the age of the pteridophytes, the world's first true land plants. These grew in myriad forms, including giant trees that flourished in the humid tropical environment of the time, many associated with extensive swamps that provided perfect habitat. Club mosses, horsetails and spike mosses also grew large. Climate change saw the end of many of these wondrous plants but some of the oldest elements of the world's ferns are well represented in the Gondwana forests, including a crepe fern and many species of ground and tree ferns. The age of conifers is also well represented by the most ancient and primitive species, including *Araucaria* and *Podocarpus*. The World Heritage area has an outstanding representation of the age of flowering plants, with plants like southern beech and numerous species in the primitive families of Winteraceae, Trimeniaceae, Annonaceae, Monimiaceae and Lauraceae endemic to the area. These species show the connections with ancient Gondwana and the subsequent landmasses of South America and New Zealand, where these plants also grow.

Looking south to Mount Warning in Wollumbin National Park from Best of All Lookout, Springbrook National Park, New South Wales.

The forests themselves are distinctly different, and provide unique ecosystems and experiences for visitors. The plant forms, diversity and colours of the area's subtropical rainforests are similar to those of the tropical rainforests of northern Queensland, and include epiphytic orchids and numerous ferns. This contrasts with the tall, stately trunks of the warm temperate rainforests, within which grow different species. The cool temperate rainforests at high altitude (for Australia) are dominated by southern beech and frequently clothed in mist, providing a 'cathedral calmness'. The adjacent surrounding vegetation—including eucalypt woodland, wet sclerophyll forest and heath—is in sharp contrast, with the transition usually occurring over a very short distance. The geomorphic features—tall mountains with dramatic cliffs and vistas—enhance the scenic qualities of these major reserves and create numerous opportunities for appreciation of the forests and the other vegetation. Hikers can enjoy the spectacular scenic views and encounter the wonderful wildlife along the walking tracks in Lamington National Park or in the Border Ranges National Park (in particular at Pinnacle Lookout).

These rainforests occupy a biogeographic region in Australia known as the Macleay–McPherson Overlap, the area on the east coast that has both cool temperate and warmer subtropical species of plants and animals. The forests are much richer than might otherwise be expected because of this overlap, meaning some parks host both temperate and subtropical species (though at opposite ends of the park). There are almost as many frog species in the Gondwana rainforests ([illegible] as in the tropical rainforests

As the ancient supercontinent Gondwana broke up, Australia slowly drifted north, a movement that is still occurring today at a rate of just under seven centimetres each year.

of the Wet Tropics of Queensland World Heritage Area (54). There are several species endemic to this area, including the fascinating hip-pocket frog (or marsupial frog), whose males rear the tadpoles in skin pouches. Similarly, there are many reptiles confined to these reserves. One of these is the land mullet, one of the largest skinks in the world, growing to 60 centimetres in length.

We now know that songbirds, one of the greatest gifts to nature across the world, originated in Australia. Two species of lyrebird, perhaps the greatest songsters of all birds, live in this World Heritage area. The Albert's lyrebird is confined to this region and each winter breeding season the males put together a repertoire of numerous calls they mimic from other species. Each male will call from his preferred cleared area and use both the call and a display of tail feathers to attract females, who do the nest building and chick raising. The slightly larger superb lyrebird also lives here and is even more impressive in the diversity of calls it mimics, including more than 20 other bird species and many other forest sounds, such as chainsaws, camera shutters and car alarms. Other ancient songbird lineages represented here include scrub-birds, treecreepers, bowerbirds and catbirds as well as a single endemic bird of paradise species, the paradise riflebird. Visitors to Lamington National Park rarely miss seeing the beautiful black and yellow male regent bowerbirds.

OPPOSITE: The striking regent bowerbirds are a highlight of any visit to Lamington National Park in Queensland.

ABOVE: The hip-pocket frog is endemic to the Gondwana Rainforests of Australia World Heritage Area—it is unique to this location.

I have visited many, but not all, of these wonderful 'islands' of rainforest. Each separate site offers variations on the connecting theme of this World Heritage area, ancient rainforest. Lamington National Park, on the border of Queensland and New South Wales, is perhaps my favourite. One August, I spent some time walking along the forest tracks enjoying the many features, especially the spectacular trees, but also looking for some of the endemic birds. I was especially delighted to see three pairs of the Australian logrunner, a relict species now confined to this area. There are two other species of logrunner, one confined to the wet tropics (the chowchilla) and the other only found in the New Guinea highlands (the Papuan logrunner).

All share the habit of scratching the leaf litter on the forest floor to expose insects and other arthropods that they eat. They have perfectly designed legs to enable lateral scratching, creating a very efficient search capacity. I watched each pair undertake their searches, leaving behind small, bare, saucer-shaped areas, from which they had gleaned their food. Also present were eastern whipbird pairs, which similarly fossick in the leaf litter for food. I noticed several wonga pigeons feeding on the ground being surprisingly quiet. Another exciting observation was a single Albert's lyrebird slowly walking through the forest. On this occasion, it did not sing. All of this was within a short distance of roads and car parks, reminding me that with little effort people are able to enjoy the unique wildlife in our World Heritage areas.

Many forms of wildlife hide in the rainforests and it usually takes a degree of luck to find them. Among these are freshwater crayfish, which occupy the many streams that drain the higher outcrops and flow across the valleys to the coast. Each rainforest fragment holds one or more of these endemic Gondwanan creatures, usually feeding along the stream bed or hiding in the pebbles and river stones. One species, the blue spiny crayfish, is confined to the Lamington Plateau, usually above 300 metres' elevation. After rain, it will roam the forest floor, feeding on leaf litter and rotting detritus. It grows to about 13 centimetres in length and is often bright blue in colour.

One of the most fascinating and ancient life forms is the velvet worm (or peripatus), a member of the Onychophora phylum with origins at the start of the Cambrian. These astonishing creatures are related to arthropods and look somewhat like a large caterpillar, two to four centimetres long but with about 14 to 16 pairs of legs. They have a distinctly velvet appearance (often dull blue in colour) and are solitary roamers of the forest. The general name of the group of peripatus is derived from the word peripatetic—one who walks or travels about. Australia is richly endowed with more than 70 species of velvet worm, many of which are endemic to the rainforests in this World Heritage area. They are voracious and active carnivores, squirting slime to capture prey such as spiders, termites and woodlice. The victim is then cut up with a pair of sickle-shaped jaws, softened by digestive saliva and swallowed. The occasional encounter between a male and female leads to packets of sperm passing to the female, who may sometimes store them for later use. As with the crayfish, each of the 'islands' of rainforest will usually have an endemic species not found in any other patch. This is a consequence of the prolonged isolation of these patches as the Australian climate became drier. The modern velvet worm provides a link between our Gondwana Rainforests of Australia World Heritage Area and the Canadian Rocky Mountain Parks World Heritage Area.

A KEY FEATURE OF THIS MAGNIFICENT SERIES OF RAINFOREST RESERVES IS THAT EACH OF THE MAJOR PATCHES DIFFERS FROM THE OTHERS, PARTLY A FUNCTION OF THE ONGOING BIOLOGICAL PROCESSES, INCLUDING EVOLUTION, THAT HAVE CONTINUED SINCE THE FRAGMENTS BECAME SEPARATED.

The latter is home of the Burgess Shale deposits from which the ancient ancestor of the Onychophora was first discovered 100 years ago. It was originally classified as a genus called *Hallucigenia* because its apparent weirdness was thought to be the result of hallucinations. At the time, it seemed like nothing on earth, a strange creature, though quite at home in shallow seas 500 million years ago. Its proper taxonomic classification was only uncovered in the 1980s.

A key feature of this magnificent series of rainforest reserves is that each of the major patches differs from the others, partly a function of the ongoing biological processes, including evolution, that have continued since the fragments became separated. This allows scientists to develop greater understanding of the processes that have shaped Australian flora and fauna since the landmass' original break with Gondwana and slow drift north.

PREVIOUS PAGES: The rainforest of the Border Ranges National Park, just south of the New South Wales–Queensland border.

OPPOSITE: The blue spiny crayfish, endemic to Lamington National Park, Queensland.

If we don't act now to save our rainforests, then this is all we'll have left. Pictures and posters of them.

Australia has already destroyed a great proportion of its natural rainforests. In N.S.W., it's estimated that less than one-third of the area of rainforest existing at the time of European settlement remains. The total areas were not large to begin with. If we keep destroying them at this rate, we could lose them forever.

So our children may never experience the unique beauty, and tranquillity of an untouched rainforest. They may never see the many different species of plants and animals that depend on our rainforests for their survival.

But there is one last hope. You.

If you write to the Premier, or the Minister for Planning and Environment at the address below, we can save our rainforests. There are many good reasons why we should.

There's more to rainforests than timber.

In fact, rainforests are the most complex communities on earth. Each one is a world of inter-dependent plants and animals. From the towering trees that soar upwards towards the canopy, to the orchids and ferns below. Removing one species often affects many others, sometimes fatally.

Many of the animals are not found anywhere else, so if we continue to destroy our rainforests much of this wildlife will disappear too.

We know very little about the great variety of plant and animal life in our rainforests. If we lose them we will also lose a valuable resource for scientific research into ecology, genetics and many other subjects. Gone, before we could learn even a little about them.

In some areas rainforests protect our water supplies by stabilising the soil. They help prevent excess erosion, muddied rivers and more frequent flooding.

Last, but certainly not least, is the fact that rainforests provide us with a valuable resource for recreation. They are beautiful, airy, moist and silent. A relaxing contrast to the smog, pace, tension and general hustle and bustle of city life. A contrast few of us will ever enjoy unless we stop the destruction of our rainforests.

There's very little left in its natural state.

Rainforests all over the world are being drastically reduced by development of one form or another. Australia is no exception. In New South Wales, they are still being utilised for timber production.

The National Trust and the State's conservation movement believe that this situation should not be allowed to continue.

Developed countries like our's can afford to keep their small areas of rainforests in an untouched state. Whilst an immediate cessation of logging is not without its problems, it is necessary if we are to retain the remaining rainforests in their natural state.

Logging is destroying our natural rainforests.

And what for?

Plywood and veneers which can also be produced from trees such as flooded gums and pines. Trees which are far more abundant than rainforest species. Indeed, rainforests as we know them today may never be seen again if they continue to be "selectively logged". "Selective logging" means cutting down the tallest and straightest trees which are centuries old and one of the rainforest's most distinctive features.

Selective logging has many adverse effects. The most obvious is the reduction in the diversity of plant and animal life. Animals are killed directly through the logging process and indirectly by the destruction of their habitats.

Logging increases water runoff which in turn accelerates soil erosion, transports vital nutrients out of rainforests and muddies our water supply. With the rainforest floor more exposed to the elements, less moisture is retained in them and they are far more susceptible to destruction by fire. The environment is so changed that many plants cannot grow back.

The effects of logging all add up to a drastic reduction of the natural beauty of our rainforests.

Their loss is too high a price to pay for the limited timber supplied, especially when there are alternatives.

Rainforests and jobs.

There is so little of our rainforests left that it is only a matter of time before timber supplies from them are exhausted. The National Trust and the conservation movement believe attention must be given immediately to the creation of jobs for those likely to be put out of work by a halt to rainforest logging.

The Trust advocates:

(a) An immediate assessment of eucalypts as an alternative to rainforest timber. This may enable the mills currently using rainforest species to switch to hardwoods.

(b) That a reafforestation scheme be introduced to provide timber for the mills in perpetuity. This would allow the harvesting of mature eucalypts now without reducing the timber resources available for use in 35-40 years.

(c) The provision of financial assistance to mills to purchase new equipment so that eucalypts can be used.

Here's how you can help.

It's simple and it won't take long.

All you have to do is write to one of the people listed below saying why we should immediately stop the destruction of our few remaining rainforests.

But do it now!

Otherwise they will be gone within twelve years. You see, you really are their last hope.

Write to:

The Premier of N.S.W.	The Minister for Planning & Environment
Hon. N. K. Wran, Q.C.	Hon. E. L. Bedford,
c/- Parliament House,	c/- Parliament House,
SYDNEY, N.S.W. 2000	SYDNEY, N.S.W. 2000

If you would like to help with a donation, then please send it to:

Rainforest Campaign Fund, The National Trust of Australia (NSW), Observatory Hill, SYDNEY, N.S.W. 2000

This poster has been created to provide you with the information you need to put the case for rainforest conservation to your local member, other Ministers of State Parliament and the media. The National Trust and the sponsors below hope you will use it.

LIST OF SPONSORS:

Australian Conservation Foundation
Colong Committee
Federation of Bushwalking Clubs of New South Wales
Friends of the Earth
Greenpeace
International Union for the Conservation of Nature and Natural Resources (Austalian Committee)
National Parks Association of New South Wales (Sydney Branch Committee)
Nature Conservation Council of New South Wales
Society for Growing Australian Plants (New South Wales Regional Committee)
The Amalgamated Metal Workers' and Shipwrights' Union (N.S.W. State Council)
Total Environment Centre
Wildlife Preservation Society of Australia (New South Wales Regional Committee)

THE NATIONAL TRUST OF AUSTRALIA (NSW)

Color photograph: Howard Hughes. Black & White photographs: Michael Morecome, courtesy "Australia the Wild Continent" by Landsdowne Press.

PATH TO WORLD HERITAGE

DURING THE 1970S, a combination of new science and changing community values led to a great shift in perceptions about rainforests. The view that a rainforest was simply a resource for logging or clearing gave way to a growing focus on conservation. Scientists began to understand the significance of the unique Australian rainforests, providing the basis for stronger community concern about their loss. After a series of well-publicised public debates and formal inquiries, the New South Wales Government adopted a new rainforest policy in 1982 that resulted in major additions to the park and reserve system.

Following recognition of the forests' significance and critical need for their conservation, the New South Wales Government supported the idea of nominating a series of these rainforest reserves for World Heritage listing. A total of 17 separate sites of largely undisturbed rainforest, all in some state of conservation status, mostly national parks, were nominated in 1984. At the time of the IUCN evaluation, it was noted that the 'documentation for the nomination ... is the most comprehensive ever received for a nominated natural property'. The first listing, in 1986, did not include any areas in Queensland, despite the recognition that these were of global significance. It was a purely political decision by the then Queensland Government not to allow any Queensland sites to be included. The name given to the World Heritage area at that time was therefore Australian East Coast Subtropical and Temperate Rainforest Parks.

In 1992, following a change of government in Queensland, a new nomination was prepared by the Australian Government that included a large number of Queensland sites. Expanding the area to the boundaries recognised today, this new proposal was warmly welcomed by UNESCO, which listed the area in 1994 as the Central Eastern Rainforest Reserves of Australia (CERRA).

In 2007, UNESCO agreed to Australia's request to change the name again to Gondwana Rainforests of Australia, which Australia felt better reflected the various qualities of the site. There is now a cooperative management arrangement between the New South Wales, Queensland and Commonwealth governments.

The 2019–2020 bushfires created severe impacts on many parts of this site. The 'unprecedented scale of fires that affected this property, which contains sensitive ecosystems not adapted to fire, is of utmost concern' to UNESCO. Three of the national parks had more than 90 per cent of their area affected. The concern remains but there have been encouraging signs of recovery and efforts made to better understand the impacts of the fires and improve emergency management. Extreme bushfires are part of the serious effects of climate change that threaten several Australian World Heritage properties. Additional concerns arise from the presence of the invasive and destructive plant fungus myrtle rust. Australia has committed to targeted recovery and resilience work in this property as well as a broader commitment to the global effort to reduce greenhouse gas emissions.

OPPOSITE: The National Trust of Australia produced posters such as these in the 1970s. They called on Australians to plead with the New South Wales Government to stop logging in New South Wales rainforests.

ABOVE: A landscape view of the eucalypt forests and woodlands of the northern section of Main Range National Park after the 2019–20 bushfires reveals extensive damage.

ULUṞU-KATA TJUṮA NATIONAL PARK

Like many parts of the ancient continent of Australia, the story of Uluṟu-Kata Tjuṯa began in the far mists of time, before land plants or animals. Over the course of a billion years, through a multitude of changes, there eventually emerged these two astonishing outback features: the stark beauty of Uluṟu and its companion Kata Tjuṯa. At the centre of the continent, remote from almost the entire population, they capture the imagination and wonder of Australians and people the world over. They are also a central focus for the people who have made the red centre home for tens of thousands of years—the Aṉangu, who claim occupation of these lands since time began and whose past and present adds much more to this outstanding place. There are few other places on earth where the interwoven story of people and place can still be seen.

> *TJUKURPA* EXPLAINS THE ORIGINS OF ALL LIFE AND LANDSCAPE IN THIS SUPERB DESERT ENVIRONMENT. IT ENCOMPASSES 'RELIGION, LAW AND MORAL SYSTEMS, DEFINING THE RELATIONSHIP BETWEEN PEOPLE, PLANTS, ANIMALS AND THE PHYSICAL FEATURES OF THE LAND'.

RISING ABRUPTLY FROM the vast sandy plains of central Australia is an instantly recognisable red rock. Along with the equally spectacular Kata Tjuta rock domes, Uluru forms the focus of this World Heritage area, which covers 132,500 hectares of arid ecosystems. These features are also at the heart of one of the world's oldest human societies, the Pitjantjatjara and Yankunytjatjara people—known as the Anangu—whose traditional lands cover the entire World Heritage area. The intimate relationship between the Indigenous people and the environment is a core element of this site. Their lives are underpinned by *Tjukurpa*, which explains the origins of all life and landscape in this superb desert environment. AIATSIS describes *Tjukurpa* as encompassing 'religion, law and moral systems, defining the relationship between people, plants, animals and the physical features of the land'.

Uluru is a huge red sandstone rock, with relatively smooth eroded surfaces, almost ten kilometres in circumference at its base and rising almost 350 metres above the surrounding plain. But what we see is a small part of the whole, as the rock (arkose) that forms Uluru extends deep beneath the earth, compressed, warped and uplifted over the hundreds of millions of years since it was formed. The geology is complex but began about 900 million years ago with the formation of the Amadeus Basin, a depression in the earth's crust that was part of the huge inland sea that covered most of central Australia. Life at that time was limited to bacteria and algae, and erosion was a powerful force. Despite very great uplift—part of the mountain building that formed the Petermann Ranges—the rocks were eroded to form extensive alluvial fans of great depth. Ongoing burial, compression and another round of uplift and folding was followed by significant erosion during a wetter period around the time the dinosaurs became extinct (65 million years ago). Then, the Australian continent was still connected to Antarctica as part of Gondwana. As Australia drifted north, the climate gradually changed and over the past 500,000 years has been so dry that windblown sand featured in the formation of the present-day sand plain and dunes.

Uluru has a series of erosion-resistant ridges and intervening valleys, some quite deeply eroded into gorges. In places, the pathways of water run-off are obvious even when there has been no rain, but during rainy periods waterfalls form as the water rushes down the slopes. Some feature erosion pools that fill with water before the flow continues all the way to the base, splashing into long-lived pools. Along the slopes, erosion has created caverns and caves, some perhaps relicts of when the rock was submerged beneath the land surface. Rock paintings can be found in some of the caves around the base, part of the enduring cultural tradition of the Anangu.

The Anangu closely follow and nurture *Tjukurpa* through ceremonies, songs, dances and other activities. The stories relate the way ancient heroes travelled across the land creating features and traditions, providing guidelines for the appropriate way of living, such as resource management, hunting and gathering rules and social relationships. One of the significant ancestors is Mala (the rufous hare-wallaby), about whom there are many stories. The law is partly reflected in the many discrete elements of the landscape

PREVIOUS PAGES: The stark beauty of Uluru, a burnished red monolith amid a flat desert plain, is recognised all over the world.

OPPOSITE: Up close, visitors can see the many cips and hollows on Uluru that have been eroded over hundreds of thousands of years.

wrought by the ancestral beings: the pockmarks on Uluṟu; the permanent waterhole at Muṯitjulu; the rocks that were the eggs of Kuniya (the woma python); and many other places that similarly offer physical evidence of the creation stories of the past. *Tjukurpa* also guides how to patch-burn the country, how to maintain waterholes and where and when to find essential resources for living. This traditional knowledge of the Aṉangu has become an essential component in managing the World Heritage area today. In responding to bushfires, for example, traditional Indigenous fire management strategies have been adopted, which have successfully prevented the larger, disastrous fires that have occurred elsewhere in the region since European settlement. At the time of its revised World Heritage inscription in 1994, Uluṟu-Kata Tjuṯa was just the second site in the world to be recognised as a cultural landscape, a testament to the ongoing living connections between the Aṉangu and the environment.

About 32 kilometres to the west of Uluṟu is the group of 36 steep-sided domes that form Kata Tjuṯa. These have a similar geological history to Uluṟu but are formed from conglomerate, a type of sedimentary rock that contains more pebbles, rounded stones and boulders embedded in the original sand and mud matrix, all fused from the rock-forming processes. On close inspection, it is easy to see the difference between the rocks at Uluṟu and those at Kata Tjuṯa. As with Uluṟu, the broad external appearance is smooth and rounded, as aeons of rain and wind have sculpted the surfaces. Kata Tjuṯa did not assume a monolithic shape like Uluṟu, but was rather eroded into a series of domes separated by canyons and passages.

OPPOSITE: Walpa (Olga) Gorge at Kata Tjuṯa

ABOVE: Kata Tjuṯa.

The landscape is an outstanding example of central Australian desert, dominated by low shrubs, herbs and spinifex with occasional scattered desert oaks breaking up the red sandy dunes and plains. There are occasional patches of mulga woodlands and near the base of Uluṟu itself grow larger bloodwoods, wattles and grasses. Over 150 species of bird have been recorded in the area, many well adapted to desert life and several featuring in Aṉangu stories. The crested bellbird has a distinctive, recurring, frequently heard call and is often difficult to locate, as the bird seems to practise ventriloquism. It starts its liquid call softly then intensifies—noted Australian ornithologist Graham Pizzey described one call as repetition of 'pan-pan panella'. The Aṉangu name for this bird is *panpanpalala*, an excellent rendition of its common call. This species features in a *Tjukurpa* story that demonstrates the shame of greed and dishonesty.

Rare mammals feature in the park and are part of its World Heritage significance, but sadly a number of species have become extinct since European settlement. The rufous hare-wallaby (*mala*) used to be abundant and widespread throughout central Australia but is now extinct in the wild on the mainland, surviving only in captive populations. The main factor in this extinction was the introduction of feral animals (cats and foxes) and the change in fire regime after Europeans ended Indigenous land management practices. Since regaining ownership of the land and co-managing the site, the Aṉangu have reintroduced the mosaic-style patch burning that supported the survival of small animals by preserving sections of habitat during burning. A program of reintroduction is underway, with captive breeding within the park. The Aṉangu are hopeful that eventually the *mala* will once more roam free in the World Heritage area.

The astonishing marsupial mole also lives here but is rarely seen, as it spends most of its time underground where it 'swims' through the sand about 20 centimetres below the surface. Little is known about this golden-furred mammal whose pelts were once traded. Its conservation status is unclear because of this lack of information, but there is a program underway to find out more. This small marsupial has several adaptations for its underground lifestyle, including a backward facing pouch, vestigial eyes, no external ears and specialised hands for scooping sand. It feeds on ants, termites and beetle larvae. Other rarely seen and threatened species in the park are the mulgara (a small marsupial carnivore), greater bilby, golden bandicoot and black-footed rock-wallaby.

The Aṉangu describe four main ecological zones in the park and in each of these live different species of bird. *Puli* refers

PREVIOUS PAGES: Vertical striations on Uluṟu are caused by rainwater gushing down the sides of the rock.

ABOVE: Traditional Owners work closely with park rangers, Muṯitjulu Community Rangers, Tjakuṟa Rangers and the Central Land Council to conduct cultural burns throughout the cooler months of the year.

to the rock faces and vegetated hill slopes where falcons and kestrels may be seen. On the rocky slopes around Kata Tjuṯa, among the spinifex (*tjanpi*), live dusky grasswrens, secretive but sometimes heard calling and seen running among the rocks, bouncing over the ground. Woodland and shrubs (*puṯi*) and patches of mulga (*puṯi-wanari*) provide habitat for another suite of species including honeyeaters, babblers and parrots. The open grasslands and dune areas (*tali* and *pila*) are the most widespread environments and are key habitat for crimson chat and pink cockatoos.

The two outcrops both feature stunning variations in colour. As Uluṟu reflects light at different times of the day, the colours vary enormously; the familiar brilliant red shades are most obvious early in the morning or late in the day. Rare periods of rain highlight the play of light and colour, enhanced by moving water on the rocks and occasional rainbows. Images of Uluṟu at sunset are internationally recognised and are much sought after by hundreds of thousands of visitors each year. Similar spectacles occur at Kata Tjuṯa, its highly varied profile adding yet another dimension.

The small carnivorous mulgara is one of several threatened species in Uluṟu-Kata Tjuṯa National Park.

My first visit to Uluṟu-Kata Tjuṯa was in 1988, part of a bicentennial trip with my family across Australia from Townsville to Perth through the central desert. As we approached the national park we could hardly believe the amazing sight of this rock, it so dominated the landscape. Even from a distance it was impressive; as we drew closer the sense of awe magnified as we began to appreciate its astonishing dimensions. We had seen photographs of course, but to be standing near its base gazing at it in real life was remarkable. Later, we were able to enjoy the beautiful desert oaks, the patches of colourful flowers, the desert roses and the ubiquitous spinifex.

That night we lay on our backs and gazed at the brilliant stars, so much clearer and more impressive than we had seen them before. The next day, we walked a track around the base of Uluṟu, seeing rock paintings. In one section, we explored a gorge, filled with trees and shrubs, where we heard the grey-headed honeyeater calling. Later, we explored Kata Tjuṯa and its walking trails, marvelling at the differences and similarities between the two outcrops. While the sun dropped lower in the sky, we joined many others to watch the light changing colour on Uluṟu. It was remarkable how quickly the colour changed, and the range of reds seemed impossible. Those experiences have stayed with me ever since.

Aṉangu storytelling and management often focuses on various reptiles—there are 77 species in the national park area. The perentie (named Ngiṉṯaka in Dreaming stories) is Australia's largest lizard, growing up to 2.5 metres in length and weighing up to 20 kilograms, and is found throughout the park. They are keen hunters, climbing trees to prey on birds and their young. Perenties are also an important food source for the Aṉangu. The king brown, or mulga, snake (Liru) is highly venomous and an important part of *Tjukurpa*. Growing to three metres in length, it is often aggressive and should be avoided. By contrast, the woma python (Kuniya) is a non-venomous snake that hunts small mammals, birds and lizards. One reptile species often seen in the daytime is the beautiful thorny devil (*ngiyaṟi*), which feeds on ants. It can sometimes be seen stationed near an ant trail, flicking the occasional ant into its mouth. While this small dragon looks fierce, it is quite harmless and slow moving, with a strange rocking motion.

Over 400 species of plants grow in the park, many crucial to the survival of traditional lifestyles. The Aṉangu use them for food, medicine, tools and weapons. Given the low rainfall (on average 300 millimetres each year), it is not surprising that many species of wattle are found here. They are valuable trees and shrubs: the Aṉangu eat their seeds and also collect large moth larvae found in the roots of some shrubs (witchetty grubs or *maku*). The honey grevillea has large yellow flowers, rich in thick, honey-like nectar that is often sucked directly from the plant. There are a number of plants from which seeds are harvested to be ground into flour. Others provide root tubers that are cooked in the fire. The once widespread and valued desert quandong is now vulnerable due to destruction by feral camels. The site's management team is returning the landscape to its condition before the introduction of feral and invasive species.

PREVIOUS PAGES: Both the Sturt's desert rose and the grey-headed honeyeater are easy to spot around Uluṟu and Kata Tjuṯa.

ABOVE: The thorny devil (*ngiyaṟi*).

PATH TO WORLD HERITAGE

THE PATH TO THIS SITE'S INSCRIPTION began with the fight for Indigenous land rights. The South Australian Government declared the area around Uluṟu and Kata Tjuṯa an Indigenous reserve in 1920. By the late 1950s, however, it had taken the land from its Traditional Owners, declaring the Ayers Rock–Mount Olga National Park in response to tourist interest in the outcrops. The title to the national park was held by the federal government, although for the eight years before handback it was run by the Northern Territory Government. As early as 1971, there were discussions between the government and the Aṉangu, who wished to reclaim their land. Despite various political and legal barriers, the land was returned to the Traditional Owners, but with an agreement that the Aṉangu would lease it

Professor Derek Ovington (Director, National Parks and Wildlife Service) and Elder Reggie Uluru hold up copies of a leaseback agreement at a ceremony officially handing back the land of the Uluṟu-Kata Tjuṯa National Park to its Traditional Owners in 1985.

back to the Australian Government (now Parks Australia) for management as a national park. On 26 October 1985, Governor-General Sir Ninian Stephen handed the land back to the Aṉangu. A board of management was created, with an Aṉangu chair (Yami Lester) and majority. This was the first example of joint management of a national park in Australia.

The initial nomination for World Heritage was under criteria that recognised the interaction between people and nature, as well as the site's natural beauty and the geological processes that created the formation. Later, when UNESCO modified the criteria, the site was renominated under two cultural heritage criteria as well as natural heritage criteria and it was inscribed as a cultural landscape, only the second in the world to be listed.

Uluṟu-Kata Tjuṯa sets a high standard for joint management. The traditional knowledge of the Aṉangu is combined with Western scientific knowledge in creating a deliberate path to recovery of environments that were modified through grazing, mining, feral animals and inappropriate fire management. In addition, Traditional Owners are employed in tourism and conservation activities, enabling them to further manage their own Country. One recent outcome of the shared vision for the area is the release of captive-reared *mala*, previously locally extinct, back into the country. It is with great joy that the Aṉangu welcome the return of the *mala* to the World Heritage area.

In 2021, UNESCO accepted a retrospective statement of Outstanding Universal Value for Uluṟu-Kata Tjuṯa World Heritage Area in which two paragraphs about the significance of the place are written in the Aṉangu language. This is first time UNESCO has included any Indigenous language associated with a World Heritage site and reflects the importance of Indigenous culture for this Cultural Landscape. In a similar mark of respect for Aṉangu culture, the decision was made by the management to no longer allow the climbing of Uluṟu.

A large camel population in central Australia has rendered many desert species vulnerable to extinction.

WET TROPICS OF QUEENSLAND

Tropical rainforests display exceptional plant growth and distinctive adaptations to continuously moist environments. A defining quality of such forests is their closed canopy, which provides a microclimate within the forest that is moist with consistently moderate temperature and relatively low light. These conditions favour diverse, lush plant growth and force a constant search for light. Generally, tropical rainforests are evergreen; rich in thick-stemmed woody lianes and other vines, epiphytic orchids and ferns; at least 30 metres high (but often much taller); and teeming with plants of large leaf size. Typically, the forest has three main layers: the upper emergent layer, where the tallest trees emerge from the canopy and are separate from each other; the continuous canopy, where the top branches of trees and the uppermost parts of lianes and vines intermingle; and the understorey of juvenile trees of the same species as the taller trees, waiting for their chance to grow in response to canopy gaps.

THEIR DIFFERENCE FROM OTHER FORESTS is often what makes tropical rainforest adaptations so visible; the tall trees, with their huge, spectacularly buttressed trunks, are instantly recognisable. Numerous species display cauliflory—flowers growing directly on the trunk. In search of light, many plants wind their way up others to find a place in the canopy, others produce very large leaves, while some stop growing altogether while seedlings and wait until a natural canopy gap occurs to trigger a speedy upward growth (known as seedling stillstand). Other adaptations commonly include drip tips on leaves to help shed rain rainwater; juvenile leaves that are large to absorb more light—though generally lacking in chlorophyll—and often richly coloured, from white to brilliant red or purple; and protective devices against predators, usually strong toxins or painful hairs and spines. Tropical rainforests have an extremely diverse range of plant species, with typically more than one hundred species of tree in a single hectare. It is sometimes said that in tropical rainforests 'rare plants are common and common plants are rare'. Matching this amazing botanical diversity is a huge array of animals with numerous adaptations to life in this moisture-laden environment.

The magnificent Wet Tropics of Queensland World Heritage Area rises majestically from the coastal lowlands to the tops of the highest mountains in the state. Extending 450 kilometres along the north-eastern coast, the Wet Tropics capture one of the last refuges of the ancient rainforests of Gondwana. Many of the plant species that grow here are sometimes referred to as 'green dinosaurs', with their origins in the far distant past when the continents of Australia and Antarctica were joined and covered by rainforest ecosystems. The World Heritage area—some 893,453 hectares—covers about 80 per cent of the area of tropical rainforest present at the time of European settlement. It represents one of the most intact areas of ancient tropical rainforest on earth. Some parts of the area have been logged for timber in the past, but recovery is advanced. Much of the lowland rainforest was lost due to clearing for agriculture and so that particular ecosystem is poorly represented now. Nevertheless, there remains an almost complete suite of habitats in the area, along with an extensive range of plants and animals that are uniquely adapted to this diverse environment.

This World Heritage area has a remarkable variety of flora and fauna. Over 3,300 species of vascular plants from 224 different families occur here, with most of the ancient groups well represented. Some 44 plant genera and 576 species are endemic. Primitive plants—those largely unchanged over millions of years—are abundant, including 15 out of the 26 primitive lineages of flowering plants in the world, the highest concentration of ancient and relict flowering plants anywhere. The Wet Tropics are also home to 65 per cent of Australia's fern species; 21 per cent of cycad species; 37 per cent of conifer species; 30 per cent of orchid species; and a total of 13 per cent of all vascular species. Together, these provide insight into the evolution of rainforests over millions of years as Australia drifted north into drier environments and its rainforests became increasingly fragmented and isolated.

The forests are also home to an astonishing array of vertebrate animals, at least 107 mammal species (11 endemic); 368 bird species (11 endemic); 113 reptile species (24 endemic); and 51 amphibians (22 endemic). Although the Wet Tropics World Heritage Area is a tiny part of Australia (less than 0.2 per cent of the landmass), within it can be found 30 per cent of Australia's marsupial species; 48 per cent of bat species; 25 per cent of native rodents; 40 per cent of bird species; 30 per cent of frog species; and 20 per cent of reptile species. Colourful butterflies are a noticeable feature of the forests: no less than 60 per cent of all Australian butterfly species are found here. Australia's two species of tree-kangaroo—Lumholtz's and Bennett's—are endemic to the wet tropical rainforests, along with a number

PREVIOUS PAGES: Noah Creek drains water from the Wet Tropics World Heritage Area into the fringing coral reefs of the Great Barrier Reef World Heritage Area.

OPPOSITE: *Stockwellia quadrifida* on the western flank of Bartle Frere: a giant ancient relative of the eucalypt, one of only 400 left.

of ringtail possum species and the musky rat-kangaroo. The endangered northern bettong and mahogany glider are confined to even smaller sections of habitat within the forests, but continue to survive, adding to the abundance of species here. In addition to the tropical rainforest, there are a number of other ecosystems in the World Heritage area, including the wet sclerophyll forests of the western margins. Here, other species of plants and animals add significant diversity to the site, including several threatened species, such as the yellow-bellied glider.

Within these tropical rainforests also live some astonishing birds, including the endemic golden bowerbird, the tooth-billed bowerbird and the Victoria's riflebird, one of Australia's four birds of paradise. Many rainforest species have well-developed songs and their calls are distinctive. Early morning here resonates with their sounds, many using calls to identify territory or connect with others. The chowchilla, for example, lives in small family parties (three to four birds), feeding on the floor of the forest. As they approach their territorial boundary, the family sings together to proclaim their ownership. They have loud distinctive calls that dominate the dawn chorus at times. The magnificent flightless southern cassowary—one of the world's largest birds—is a keystone species, responsible for the distribution of several critical rainforest tree species with exceptionally large fruit. Unfortunately, many cassowaries

OPPOSITE: In feeding, the southern cassowary spreads the seeds of a large number of rainforest tree species.

ABOVE: Aerial view of a typical rainforest canopy showing the different colours of the many species present.

die each year as they try to cross the roads that partition their habitat.

Numerous other animals have habitats restricted to these rainforests, adding further diversity to the ecology, including freshwater species, such as the platypus and rainbowfish; reptiles, such as Boyd's forest dragon; numerous endemic skinks, including the unusual prickly forest skink; Australia's largest snake, the amethystine python; and the delightful endemic relict geckoes, the chameleon gecko and the leaf-tailed gecko. Many of the frog species show interesting patterns of distribution, with some species confined to individual mountain tops. Australia's smallest frog, the scanty frog (or dainty nursery frog) is only found in the Wet Tropics, confined to rainforests in the Mount Finnigan area. Australia's largest frog—and the largest tree frog in the world—is also found in the Wet Tropics: the giant tree frog (or white-lipped tree frog). Insects are abundant, including the huge Hercules moth, the brilliant Ulysses butterfly and thousands of colourful moths and beetles. Many of these demonstrate links to the fauna of other Gondwanan fragments, including South America.

Flying fox camps are among the most interesting spectacles in the Wet Tropics area. During daylight, these keystone animals rest in camps, hanging in the canopies of their selected trees. Each branch may hold dozens of individuals and some camps number several thousand. At dusk, they set out for their feeding areas, important pollinators for many tree species. Some species, such as the little red flying-fox, will feed on blossom while others, like the spectacled flying-fox, choose fruit. They will often distribute the seeds many kilometres away from where they found them. This species is generally confined to tropical rainforest, with camps usually

OPPOSITE: Giant tree frog.

ABOVE: Spectacled flying-foxes feed on fruit at night, resting in tree-top camps during the day.

The 1,000-year-old Twin Kauri pines at Lake Barrine have been known to the Ngadjon-Jii for many generations.

within 6.5 kilometres of the foraging area. The sight of many thousands of these beautiful mammals heading out to feed is highly memorable.

As in the other two rainforest World Heritage areas in Australia—the Tasmanian Wilderness and Gondwana Rainforests—the essential importance of this site is its contribution to our understanding of southern hemisphere biological history. Plants in this area show the evolution of life on earth, from the age of ferns (pteridophytes) some 200 million years ago to the evolution of seed-bearing plants, the age of the cycads and conifers (gymnosperms) and eventually the rise of the flowering plants (angiosperms). The extraordinary array of plants that made up the rainforests of Gondwana were all rafted north as Australia broke away from Antarctica and drifted slowly to its present position, gradually encountering an increasingly dry climate that radically transformed the vegetation. From the eucalypts that once grew on the edges of rainforests eventually emerged the great woodlands and mallee communities, dry forest and grasslands that dominate much of Australia today. It is only in these few tiny refuges that we now find the relicts of what was once spread across the continent. In Tasmania, the temperate rainforests survive; in northern New South Wales and south-eastern Queensland, the warm temperate and subtropical rainforests have found refuge; and in northern Queensland, the Wet Tropics preserves a fragment of the ancient tropical rainforests.

Part of the scenic beauty of the Wet Tropics is a result of the volcanic activity that left behind beautiful crater lakes—such as Lake Eacham and Lake Barrine—as well as a series of impressive scoria cones and other features, including waterfalls. High mountains contribute variable ecosystems, habitat and the unique wildlife that live there. From a distance, these features stand out: Mount Bartle Frere and Mount Bellenden Ker, at 1,622 metres and 1,593 metres respectively, tower over Wooroonooran National Park and provide landmark skylines from the east and west.

PLANTS HERE SHOW THE EVOLUTION OF LIFE ON EARTH, FROM THE AGE OF FERNS 200 MILLION YEARS AGO TO THE EVOLUTION OF SEED-BEARING PLANTS, THE AGE OF THE CYCADS AND CONIFERS AND EVENTUALLY FLOWERING PLANTS.

Further north, Thornton Peak (1,374 metres) is a feature of the Daintree region and in the far south, Mount Spec (1,186 metres) and Mount Halifax (1,063 metres) protrude above the coastal plain. The flanks of the mountains display the dark green colour of tropical rainforest from a distance, unlike the close-up view, which is highly varied in shades of green and distinctive fresh shoot colours.

In places, the Wet Tropics area extends right down to the coast and at Cape Tribulation the boundary abuts that of the Great Barrier Reef World Heritage Area. The juxtaposition of two World Heritage areas is unusual, though there are many ecological connections between the two. The pied imperial-pigeon, for example, feeds on the fruit of many species in the tropical rainforests and can be seen foraging during the day. In the late afternoon, they fly in their thousands to the offshore islands of the Great Barrier Reef where they build their nests and raise their young. This species was once endangered by hunting but has now recovered, with both its rainforest larder and its island roosting areas protected by World Heritage listing. The monitoring of this population was very much a community project, led by Margaret Thorsborne AO and her husband, Arthur, and continuing in partnership with environmental scientists, park managers and community groups. It is a notable example of the essential importance of cooperation in World Heritage conservation efforts.

FOLLOWING PAGES: Wallaman Falls, near Ingham, is the highest single-drop waterfall in Australia; INSET: Lumholtz's tree kangaroo.

In 1975, I made a deliberate decision to relocate from the University of Western Australia in Perth to the recently formed James Cook University in Townsville. The attraction for me was the chance to study tropical rainforest. I had become fascinated by these astonishing environments all over the world, so varied in structure, so rich in species diversity, so different from the forests of the south-west of Australia. My first encounter lingers with me still. An experienced colleague, David Thomae, took me camping in the forests of the Atherton Tablelands, near Mount Hypipamee. I learnt from him that much of the wildlife is difficult to see in tropical rainforest and so you must learn to use other senses—hearing is crucial. He had found far greater accuracy and completeness from bird surveys using calls compared with traditional mist netting, with the added advantage of little impact on the birds.

We took our sleeping bags under the forest canopy and I listened while the first nocturnal species revealed themselves in sound: the croaking of frogs; the 'falling bomb' call of the sooty owl; the raucous cackle of the orange-footed scrubfowl. Repeated crunching filled the air—perhaps a white-tailed rat cracking nuts—alongside the chirping of crickets, almost drowning out the faint sound of scorpions or centipedes hunting in the leaf litter. Tree kangaroos and possums added their own noise, brushing through foliage and branches in their night-time wandering. Before dawn, the earliest birds called—pale-yellow robins, grey-headed robins—and then a sequence continuing through honeyeaters, monarchs, treecreepers, chowchillas and fruit pigeons. The dawn chorus in tropical rainforest is simply remarkable.

Some years later, when I was taking university students into the forest, it was a privilege to have a woman from the local Aboriginal community (Ngadjon-Jii) talk with us about the famous twin kauri at Lake Barrine. Visitor interpretation at the time was limited to a forestry sign that said how tall each of the twin trees was and how much timber each could produce. The Elder spoke instead of how these trees were part of her family and how they were known to her parents and her parents' parents and so on back in time. There were stories of their witness of, and survival through, amazing events over that time, long before Europeans arrived. She told of visiting these members of her family to talk with them and to share her life with them. The students were deeply affected by these revelations. Many similar ancient trees (1,000 years old) were removed in the logging era.

Throughout the Wet Tropics World Heritage area there are 18 Indigenous community groups with ongoing connections to the land, descended from the original people who first came to Australia more than 40,000 years ago. These groups have exceptional knowledge of rainforest plants and how to find, prepare and remove toxins from many of the food sources, as well as which species possess medicinal properties. They have developed ceremonies, songs and tools relating to the rainforest and its biological and geomorphological history. These tools include specialised fish traps for use in the area's fast-flowing rivers and shields made from rainforest timber, often cut from buttressed tree trunks. The unique cultural heritage of the Wet Tropics, Indigenous peoples has now been recognised and added to the National Heritage List. It is expected that the rainforest's Indigenous cultural heritage will be nominated for World Heritage listing in the future.

A poster produced by the Wet Tropics Management Authority describes the different food sources of the rainforest.

PATH TO WORLD HERITAGE

THE WET TROPICS OF QUEENSLAND World Heritage area faced one of the most challenging paths to inscription of all Australian sites. This was due not to any lack of scientific knowledge about its many outstanding attributes but as a result of significant intergovernmental political dispute over use of the forests. As early as 1966, Dr Leonard Webb from the Commonwealth Scientific and Industrial Research Organisation (CSIRO) identified the exceptional value of the rainforests in north-eastern Queensland and the Committee of Inquiry into the National Estate promoted the need to conserve these treasures. Dr Aila Keto AO, of Brisbane, established the Rainforest Conservation Society in 1982, which expanded the scientific knowledge about rainforest conservation. Many community organisations proposed further recognition of the rainforests through listing on the Register of the National Estate, a federal government program. Much of the area had already been entered into the Queensland National Park estate but substantial areas continued to be used for forestry and other areas were subdivided for real estate. The different demands on the forests—for timber by the Queensland Government and expanded conservation by various organisations and scientists—led to conflict. In 1982, the IUCN published a prospectus for future World Heritage sites and in that book the Wet Tropics of Queensland featured prominently. Much of the forest knowledge in the prospectus came from CSIRO scientists, including Leonard Webb and Geoffrey Tracey. In 1984, the Rainforest Conservation Society, at the request of the Australian Heritage Commission, produced a major study on the international significance of the tropical rainforests and the result was extremely positive. In light of the study, the federal government announced in 1987 that it would proceed with nomination of the area despite the state government's opposition.

Meanwhile, there had been enormous public controversy about the failure of the Queensland Government to protect the rainforests in the Daintree River area and the plan to bulldoze a road from Cape Tribulation to Bloomfield. This was seen as a reversal of policy issued at the World Wilderness Congress in Cairns (1980), when Premier Sir Joh Bjelke-Petersen had promised a national park. Land subdivision in the Daintree aggravated the situation; gradually, local issues became national and a long campaign involving blockades and protests ensued. By 1984 strong national media coverage was highlighting the difference between the views of the state government and many local and national groups.

The Winkel brothers, champion axemen, felling a large walnut tree on the Atherton Tablelands around 1928, long before the harmful effects of tree-felling were recognised.

STOP
crucifying
the
rainforest
NOW

The strong advocacy of national conservation and science groups supported the many active local groups and this eventually led to Prime Minister Bob Hawke's 1988 intervention to stop logging in the area nominated for World Heritage listing. Despite a legal challenge from the Queensland Government, the nomination went forward and was strongly endorsed by UNESCO, with the IUCN saying in its evaluation that 'the background for the nomination of the Wet Tropical Rainforests of North-East Australia is the most comprehensive ever received for a natural World Heritage nomination'. Over 20 scientists were engaged during the evaluation process and the IUCN inspection team spent more time than usual in assessing claims about the site. Significant negotiation over the boundary preceded the final nomination, with some adjustments reflecting compromises that subsequently meant future costs in buying back land.

The federal government made generous compensation payments to those affected by the cessation of logging and established an initially well-funded body to manage the area using a community-based system. Support from the local community for World Heritage listing was always high and increased further over time, as new industries emerged that offered a more sustainable future than logging. While the decision to nominate the area for World Heritage listing took place at a national level, it is unlikely that it would have gone ahead without the strong commitment of conservation groups and individuals, especially the regional group Cairns and Far North Environment Centre, which coordinated much of the program to achieve nomination.

There have been some attempts to maintain the scientific knowledge considered essential to understanding and managing this World Heritage site. Various rainforest research facilities have been established, but long-term and sustained funding has not been forthcoming. Some research has drawn attention to the likely serious threats to the area as a result of climate change, particularly the risk to some endemic species. There have been significant cutbacks to the budget for the management of the World Heritage area in recent times, resulting in the reduced capacity of community-based programs.

The Wet Tropics is one of our World Heritage sites likely to be severely affected by climate change impacts. In 2020, the management authority produced a climate adaptation plan to enhance protection of the site. A key focus is to protect essential nature refugia (habitats where plants and animals can survive during unfavourable environmental circumstances) for high elevation species, likely to be adversely affected by rising temperatures. Many species of plants and animals will need refugia and better landscape connectivity as conditions change. The plan requires whole of community involvement; progress is due for review in 2025.

OPPOSITE: A protester is tied to a crucifix during rallies against the destruction of the Daintree Rainforest in 1983.

ABOVE: Demonstrators parked near Old Parliament House, Canberra, around 1977.

SHARK BAY, WESTERN AUSTRALIA

Human activity is changing the earth's climate, and countries across the world are working to reverse that undesirable situation. Long ago, another species changed the entire planet. When life first appeared on earth some 3.5 billion years ago, the atmosphere was mostly carbon dioxide, quite toxic to modern life forms but suitable for an organism at the heart of the Shark Bay World Heritage Area: cyanobacteria. Although simple life forms, cyanobacteria were ultimately responsible for one of the most important global changes that the earth has undergone. Cyanobacteria are photosynthetic, using energy from the sun to extract carbon from the atmosphere and release oxygen. These were our first and only forms of life for some two billion years and gradually they oxygenated the oceans and atmosphere, making it suitable for later, more complex life forms. Cyanobacteria live in communities, some forming microbial mats that create hard structures known as stromatolites—these are the world's oldest fossils. Remarkably, they continue to survive, but today are found in just a few locations around the world. Modern stromatolites are restricted to shallow coastal waters, lakes and rivers in environments that are still harsh for other life forms. The hypersaline waters of Shark Bay in Western Australia contain the world's most diverse and abundant examples of living stromatolites.

ON THE INDIAN OCEAN COAST of Western Australia, the waters, islands and peninsulas of Shark Bay World Heritage Area cover 2,200,902 hectares and were inscribed on the World Heritage List in 1991 under all four natural criteria. Some 70 per cent of this area is marine and it was for three outstanding marine elements that the area was deemed of international significance. The first is the vast area of seagrass meadows, the largest and richest in the world. These in turn support a highly significant population of dugongs, again one of the world's largest. The third outstanding marine element is the exceptional stromatolite colonies, with their hard, dome-shaped deposits rising along the shores of Hamelin Pool and elsewhere. It was in 1956 that the first living stromatolites were discovered, at this site. Before then, it was thought that only fossil forms existed.

The islands and convoluted coastal structures with several inlets, bays and peninsulas making up this World Heritage area, have some 1,500 kilometres of coastline. These coastal environments vary significantly and include bright white, shelly beaches; long, yellow sandy beaches; and, in places, even beaches backed by bright red dunes. There are also spectacular limestone cliffs and headlands, and in places, extensive colonies of stromatolites. Many areas are well protected by offshore barrier islands (Dirk Hartog, Bernier and Dorre islands), which further enhance the coastal beauty.

The coastal sand dunes support a large array of plant species, a quarter of which are at the limit of their range. Some, such as the kangaroo paw, grow at the northern extent of the Southwest Province, the site of one of the world's richest floristic assemblages. Others, such as horse mulga, grow at the south-western limit of the Eremaean Province which covers the inland arid communities of plants. More than 50 of the 700 flowering plant species are endemic.

PREVIOUS PAGES: Stromatolites in Hamelin Pool, January 2018

LEFT: Big Lagoon, at the tip of the Peron Peninsula.

The seasonal display is particularly spectacular, including multicoloured everlastings as well as heath plants such as grevillea and hakea. One unique vegetation type is the unusual tree heath community south of Freycinet Estuary, almost all of which occurs in the World Heritage area. This is the most diverse and complex plant community of Shark Bay and includes a mix of Eremaean and Southwest province species that, combined, make up about half the endemic species in the area. Many of the shrub species are particularly large specimens, not known to grow this tall elsewhere.

An aerial view shows the great indentations of the coast into the arid brown lands of the mainland that make up Shark Bay. Blue and green ocean water fills the inlets and bays, mostly calmed by the sheltering peninsulas and three barrier islands. Behind the coast are numerous gypsum claypans—or *birridas*, as the local Indigenous community calls them—in various shapes. While mostly dry, after rainfall the dormant eggs of brine shrimp, horseshoe crabs and other species hatch and a brief thriving aquatic community attracts feeding waders. The Peron Peninsula has a very narrow section across which a fence was erected

Shark Bay has the longest wildflower season in Western Australia, lasting six months.

Shark Bay is at the far north of the regent parrot's habitat.

as part of Project Eden, enabling the capacity for eradication of introduced feral species and the reintroduction of native mammals. The greater bilby and the malleefowl have both now been successfully reintroduced but at this site persistent feral cats have inhibited the success of other species' releases.

The rich vegetation and coastal resources support over 230 species of bird, including the delightful western grasswren, largely confined to the coastal dunes and commonly seen at Monkey Mia. Some bird species, such as the regent parrot and the blue-breasted fairy-wren, are at their northern limit in Shark Bay. Numerous waders also visit the area during their summer migration from the northern hemisphere. The rich reptile and amphibian fauna includes three endemic species of 'sand-swimming' skinks. The keeled slider 'swims' through dune sands just below the surface, hunting for prey.

The extensive sheltered, warm, shallow waters are perfect for the growth of seagrasses, with 12 species here forming the world's most extensive seagrass meadow. Seagrasses

are aquatic flowering plants that form highly productive ecosystems, supporting many different animal species here, including 11,000 dugongs. This fascinating marine mammal shares a common ancestor with elephants, separating in their evolution perhaps 65 million years ago. A large herbivore (more than two metres long, weighing up to 350 kilograms), the dugong lives in tropical coastal waters throughout the Indo-Pacific region, from east Africa to Vanuatu. While the IUCN lists it as vulnerable to extinction, Australia has the largest and best-protected population in the world, ranging from Moreton Bay in Queensland through Torres Strait and around the Top End to Shark Bay. They are a long-lived but very slow-breeding species and totally dependent on the seagrass meadows. The area also provides one of Australia's most important nesting places for the endangered loggerhead turtle, as well as critical habitat for green turtles.

Whaling was an early industry at Shark Bay, supporting a whaling station at Carnarvon in the north of the World Heritage area. Initially, whales were taken from within the bay but later the hunters moved further offshore to the migration route. Between 1950 and 1962, a total of 7,352 humpback whales were killed, but the resulting scarcity of whales led to the station's closure in 1963. The population is recovering, following the end to widespread whaling, and now many humpback whales use Shark Bay as a staging post during migration. Numerous dolphins live within the bays and are a popular tourist attraction at Monkey Mia, where they interact with humans. Questions continue to be raised about the long-term viability of this activity and

OPPOSITE: Dolphins at Monkey Mia.

ABOVE: The rufous hare-wallaby is vulnerable to extinction; its small remaining population finds refuge at Shark Bay.

I first visited Shark Bay when there was little or no infrastructure there, apart from a small fish-processing factory at Denham. It was 1963, I was 19 years old, and the drawcard was the prospect of fishing in the rich waters of the bay. The abundance of fish was amazing, especially of our target, the large pink snapper, which occurs in deeper water (15 metres or so) over coral bottoms. In between the fishing, I was fascinated by the vegetation communities and the wonderful coastline, including Monkey Mia, then an undeveloped beach.

Many years later, I spent time exploring both the marine environments—as part of an investigation into ecotourism—and some of the terrestrial features. I was astonished by the stromatolites, which seemed similar to the spectacular coral reefs I was much more familiar with—the solid outcome of tiny coral polyps. More recently, I undertook a survey with a colleague and we found a butterfly not then known to science hiding away among the coastal dunes. It is highly likely that there remain more biological features waiting to be discovered.

alternative ways for human–dolphin interaction are being explored. There are many sharks and rays also highly visible within the bays, including tiger sharks, hammerhead sharks and giant manta rays. At least six species of sea snake occur in the waters, including the endemic Shark Bay seasnake.

Shark Bay is a refuge for many threatened species and the fate of five terrestrial mammals, in particular, depends on the Bernier and Dorre Islands Nature Reserve. As introduced grazing animals destroyed habitat, and cats and foxes decimated the populations of small mammals across the continent, these offshore islands—50 kilometres off Carnarvon—provided a refuge. The fascinating banded hare-wallaby, once found widely across Western Australia and South Australia, is now found only in this reserve. Similarly, the Shark Bay mouse was once found throughout the western two-thirds of Australia, including at Alice Springs. It clings on within the reserve. The rufous hare-wallaby, once widespread, is now only found here. There are ongoing attempts to undertake captive breeding and reintroduction within the Northern Territory. Burrowing bettongs, or boodies, are extinct on the mainland but the two subspecies on Bernier and Dorre islands survive and have been used to breed captive populations in Western Australia, New South Wales and South Australia. A similar story applies to the western barred bandicoot. It is fortunate indeed that these small islands preserve some of our endemic mammals and give at least a chance of reintroduction to the wild once the conditions that led to extinction have changed. A recent project has focused on Dirk Hartog Island, where vegetation is recovering from sheep and goats (now eradicated), and feral cats have been eradicated. The ambitious program is seeking to introduce up to 11 small animals that were once present in the World Heritage area. Both rufous and banded hare-wallabies have been released, and are doing well. Other missing species have also been successfully released, including dibblers and Shark Bay bandicoots, with more to follow as circumstances dictate.

Although the site is listed for its natural heritage, there is mounting evidence of an Indigenous presence dating back as far as 30,000 years. There may have been a hiatus in occupancy between 18,000 years and perhaps 6,000 years ago but, given that many potential sites occupied during that period are now underwater, the evidence is inconclusive. The Traditional Owners of this area are the Yamatji. Recent archaeological work at Boodie Cave on Barrow Island (450 kilometres north of Shark Bay) has revealed occupancy as far back as 50,000 years, perhaps the oldest so far for a coastal site. There could be more Indigenous heritage yet to be revealed.

The area was initially settled as grazing area, but it was also used for other industries, including pearling (from 1860), fishing (which peaked in the 1960s), gypsum mining and an evaporative salt industry at Useless Loop. The area's most significant industry is also the most recent: tourism. At the time of nomination in 1990, some 115,000 visitors travelled to Shark Bay each year. As in the Great Barrier Reef World Heritage Area, a number of commercial activities continue to operate, including salt mining, fishing and gypsum mining. The area's management authorities oversee these activities, ensuring all use is sustainable and that there is no damage to the outstanding universal value of the area.

OPPOSITE: Stromatolites at Shark Bay are a tangible reminder of the evolution of life on earth.

ABOVE: Solar salt fields—just one of the industries within the Shark Bay World Heritage Area.

PATH TO WORLD HERITAGE

THE CONSERVATION SIGNIFICANCE of Shark Bay to the international community was recognised in the IUCN's 1982 publication about potential World Heritage sites. There was already much scientific interest in the area, following the first record of living stromatolites in 1956. The existence of a large, stable population of dugongs, and the associated extensive seagrass beds on which they fed, was also the subject of considerable scientific interest. Much of the area was entered on the Register of the National Estate for various natural heritage reasons in 1978. Subsequently, the Western Australian Government prepared the Shark Bay Region Plan, released in 1987, and public comments included a suggestion that the area be nominated for World Heritage. In that same year, a report on the potential for World Heritage listing was prepared by the Department of Geography at the University of Western Australia for the Australian Heritage Commission.

In 1990, the Western Australian and Australian governments announced that they would be submitting a nomination to UNESCO, which they did in October that year. Some 200,000 hectares already had some form of conservation status and in 1990 the government of Western Australia purchased the 105,000-hectare Peron pastoral lease to convert into a conservation area. This subsequently became a national park and the site of important pest management operations. In 1994, Project Eden was established on the Peron Peninsula with the objective of eradicating feral species responsible for habitat degradation (goats) and predation on small animals (foxes and cats). The long-term objective was to reintroduce the most threatened species.

Following the World Heritage listing, some people continued to express concerns about the possible adverse effects on the area from the economic activities. Local threats to the property include invasive species, especially weeds; feral animals, especially cats; and potential invasive marine organisms. The lack of an overarching management plan remained a deficiency for many years. This was addressed with the release of the *Shark Bay World Heritage Property Strategic Plan: 2008–2020*, jointly prepared by the Western Australian and Australian governments. The plan identifies a series of objectives that focus on protecting the outstanding value of the site, and each of these objectives has an accompanying strategy and responsible management agency. The Western Australian Department of Environment and Conservation, for example, is responsible for investigating the potential vulnerability of species to climate change and incorporating the potential for climate change impacts into recovery plans for threatened species. The impacts on the site's outstanding seagrass beds are of particular concern, and the development of adaptive management strategies to mitigate such impacts is a high priority—as in other sites in Australia. While the aim of the plan is to better manage the area and therefore protect its heritage qualities, the reality is that plans like these depend on strong on-ground support and funding, something not always forthcoming.

OPPOSITE: A burrowing bettong on Dorre Island.

ABOVE: The Peron Peninsula conservation area is home to many threatened and endangered species.

K'GARI (FRASER ISLAND)

A million years ago, a pebble sat along a small tributary of the Hawkesbury River, somewhat west of Sydney. It rolled down the bank into a creek, perhaps dislodged by a giant marsupial. Over a few thousand years, rain flushed it downstream, where it was sometimes caught up in potholes or eddies, gradually worn, slowly descending. Eventually, it reached the coast and was swept into the sea with billions of its fellows, where the longshore current moved them north, building up sediment that allowed coral reefs to develop in the Barrier Reef. After thousands of years, the sedimentary particle, much smaller, arrived onshore again. It washed up on the beach, where it was caught by the wind and blown up and over a dune. Snagged by some uncertain feature, it was building a solid future, joining with billions of others to shape the dune higher, part of the bright white sand blow that is constantly creating and unmaking the dunes of K'gari. Imagine that: a particle that once resided in what is now the Greater Blue Mountains World Heritage Area travelled all the way north to nurture and produce the K'gari World Heritage Area. Each of the billions of grains there has a similar story to tell.

STRETCHING OVER 120 KILOMETRES parallel to the southern Queensland coast, K'gari is the largest sand island in the world. It is a place of exceptional beauty, with long white beaches fringing the massive sand dunes and freshwater lakes set among tall rainforests and heath. The tallest dunes reach 240 metres above sea level and have a long history of development extending back almost 900,000 years. The island is an amazing record of geomorphological processes and features including the oldest and most complete sequence of coastal dunes on the planet. The entire island and a fringe of marine waters are included in the World Heritage area, covering 181,851 hectares.

Erosion from the Great Escarpment along eastern Australia produced huge volumes of silica sediment, which was washed down streams and rivers and swept out to sea. The prevailing south-east winds pushed the washed and worn sands towards the shorelines. Following glacial sea-level falls, and during subsequent post-glacial rises, the sediment moved north and came to rest eventually as the various islands of Moreton Bay and further north, with the largest mass of all settling off Hervey Bay. Strong winds continued to shift and shape the dunes as they moved inland from the eastern beaches to form many different dunes. Gradually, K'gari grew, from the oldest dunes in the west to the youngest in the east. Its present structure as an island goes back perhaps 6,000 years, with the last post-glacial sea-level rise separating it from the mainland.

As the older dunes stabilised with vegetation, the system of organic matter and superfine particle accumulation beneath the surface enabled the gradual formation of lakes, including half the world's perched freshwater dune lakes. A perched lake forms when organic material like dead plants, leaves and bark collects in a depression within a sand dune. Over time, this material decomposes, forming a sealed layer over the sand underneath that then collects water.

PREVIOUS PAGES: The sand dunes at K'gari form some of the most recognisable beaches in the world.

LEFT: Sandy Cape, the northernmost point of K'gari.

More than 100 of these freshwater lakes form part of the exceptional natural beauty of K'gari, including Lake Boomanjim—at almost 200 hectares, the world's largest perched freshwater lake. The dunefields remain active and large parabolic dunes, without vegetation, migrate across the island, driven by the wind. In places, these mobile dunes impede streams and form yet another type of freshwater lake, a barrage lake, such as the beautiful Lake Wabby. The creation of coloured sands, reflecting iron oxides in the source material and sometimes stained by oxidised organic material, adds another spectacular dimension to the island.

Created entirely from sand, it might be expected that this very-low-nutrient island would be unable to support much vegetation. But, partly because of the age of many of the dunes, there has been an accumulation of nutrients and so the island has a diverse range of plants and vegetation types. Most surprising, perhaps, K'gari is the only place in the world where tall rainforests grow on sand. The forests are dominated by several species of tall tree, including satinay (some of these are 1,000 years old), kauri pine (once the focus of logging), brush box and piccabeen palm, as well as bollywood, bumpy ash and ribbonwood. As in most rainforests, there are epiphytes and ferns and numerous vines. Orchids are common and epiphytic ferns—such as crow's nest fern, staghorn and elkhorn—festoon the taller trees. The rainforests grow mainly down the central spine of the island and reach peak condition in the valleys, inter-dune corridors and along streamlines.

Where the dunes are younger, with less accumulated nutrients, there are several other vegetation types. Fringing the rainforest are tall eucalypt communities of blackbutt and tallowwood, with some bloodwood, brushbox and forest she-oak. This forest also has a dense understorey of shrubs. Closer to the coastal dunes is a more open eucalypt forest, a result of less well-developed soil and more exposure to the strong winds. There is then a transition to grassy woodlands and finally, on the eastern area of exposed fore dunes, examples of the earliest plant pioneers, including casuarina, helping to stabilise the sand, especially sprawling vines of convolvulus and scattered pigface. Each successive plant community helps to protect the plants in their lee and forestall the eruption of another sand blow. In many of the deeper sands, there are stands of heath including *Banksia integrifolia*, which also grow adjacent to other communities along streams and around lakes and swamps.

OPPOSITE: Lake Boomanjin, the world's largest perched freshwater lake.

ABOVE: Forests of satinay, and other tall tree species, dominate K'gari.

Paperbark trees are also widespread. On the sheltered western side of the island, many mangroves (ten known species) provide extensive habitat for marine species. Across all the vegetation types, there are over 600 plant species on the island, including around 50 species of fern.

Many animals here have adapted to the diverse geomorphological and floristic environment, including 350 recorded species of bird. Some of these are of high conservation significance, including migratory waders that fly from the northern hemisphere in Australia's summer months, some from Siberia, China and the Arctic Circle. These include grey-tailed tattler, lesser sand-plover, Terek sandpiper and grey plover.

Several, including the critically endangered eastern curlew, depend heavily on the area. These migratory waders are also protected as part of bilateral agreements with Japan, China and the Republic of Korea, reflecting the widespread international concern about their conservation. Another bird species of particular note is the eastern ground parrot, an endemic Australian species that is considered vulnerable in the eastern mainland states. The eastern ground parrot is a slender, long-tailed, grass-green parrot, mottled black with

K'gari is home to the rare ground parrot.

a hint of yellow; it also inhabits the Tasmanian Wilderness World Heritage area. Ground parrots are confined to low heath and feed on the seeds of associated sedges. Sightings of the ground parrot are not common, as they rarely fly and live in quite dense wallum (an ecosystem characterised by shrubland and heath, dominated by the wallum banksia). Despite their vulnerable status, they are likely to survive on K'gari, especially with appropriate fire management that ensures fires are small and patchy rather than extensive and frequent. The island supports one of the most extensive patches of surviving wallum in Queensland.

Although the boundary of the World Heritage area only extends 500 metres into the marine habitat surrounding the island, that environment adds significantly to the diversity of life. Significant seagrass beds of the Great Sandy Strait (Korrawinga) support populations of endangered dugongs, with six of the world's seven species of marine turtles also recorded in the region. Turtles are common on many of the coral reefs of the Great Barrier Reef and in some sheltered parts of the Queensland coast, including the Great Sandy Strait. The Capricorn Bunker Group of islands in the southern Great Barrier Reef World Heritage Area is one of the world's major breeding areas for turtles. Endangered loggerheads nest on K'gari at Sandy Cape, with Mon Repos supporting the largest concentration of nesting turtles on the eastern Australian mainland. The same species, but from the Indian Ocean population, breeds in Western Australia's Ningaloo Coast World Heritage Area. Migrating humpback whales visit the sheltered waters for several days at a time and there are resident dolphin populations. In the Great Sandy Strait, there are abundant fish resources and associated predators.

As noted in the nomination, the island has much archaeological, historical and spiritual heritage associated with the Traditional Owners, the Butchulla people, whose Native Title rights were recognised in 2014 and 2019. There is archaeological evidence of Indigenous people

THE ISLAND HAS MUCH ARCHAEOLOGICAL, HISTORICAL AND SPIRITUAL HERITAGE ASSOCIATED WITH THE TRADITIONAL OWNERS, THE BUTCHULLA PEOPLE, WHOSE NATIVE TITLE RIGHTS WERE RECOGNISED IN 2014 AND 2019.

on the island going back 1,500 years, although they lived in the wider region for many tens of thousands of years before then. On K'gari itself, there are large middens, demonstrating the richness of the coast in supplying food, as well as fish traps, artefact scatters, scarred trees and campsites, burial sites and ceremonial sites. K'gari holds a special place in Butchulla culture, with the island's lakes part of sacred creation stories. The Queensland Government decided, in April 2017, to change the name of the Fraser Island section of the Great Sandy National Park to the K'gari section. Official processes are underway to formally declare the island name K'gari. It is interesting to note that at the time of listing (1992), the IUCN noted that K'gari was the local Indigenous name for the island and recommended that Australia consider naming the site 'The Great Sandy/K'gari'. The implication was clear: K'gari World Heritage Area was a sufficient name without the inclusion of 'Fraser Island'.

K'gari World Heritage Area is of global significance because of the ongoing geomorphological processes that led to its creation and shape its present nature; the associated biological processes; and the exceptional beauty of the landscape and seascape. Many features are exquisite in their own right. Added to these attributes are the various human dimensions of Indigenous culture, past and present, and the experiences offered to the many visitors who enjoy the island today.

FOLLOWING PAGES: Extensive mangroves provide habitat for many marine species on K'gari.

My first visit to K'gari, many years ago, was not as a researcher, but rather as a participant in a workshop considering the relationship between protected areas and ecotourism. But given my interests, I simply had to explore the island further. As someone who had studied rainforests for a long time, I was stunned by the sight of subtropical rainforest, tall and lush, on pure sandy soils. I walked along a creek of crystal-clear water that flowed through the rainforest, its bed entirely white sand. It seemed so improbable that such a forest could exist and yet here it was: the only example in the world of a rainforest growing on sand dunes. On another occasion, I spent a day at Lake McKenzie, one of the magnificent perched freshwater lakes. The features that I recall are many, including delightful paperbarks sprouting up in places along the white sandy shore.

The water was extremely clear, the temptation impossible to resist and I enjoyed a quiet dip, screened by fringing rushes but looking across the green and blue lake surface to a distant sandy beach. I felt the sand between my toes and under my feet. Every grain had been swept here by the wind, so was quite small and fine to touch. These grains, or some of them at least, had once formed part of the Great Dividing Range in the Blue Mountains of New South Wales. I find that thought astonishing. I marvelled at the endless kilometres of surf beaches along the east coast; the patches of wallum, with terrific plant diversity; the cross-island sand blows breaking up the long, northward-trending coast. It is a very special place and World Heritage listing has brought it a better level of protection. I felt admiration and gratitude towards John Sinclair, a leading conservationist who worked his entire life, against significant odds, to protect this treasure.

PATH TO WORLD HERITAGE

IN 1836, THE TRADING VESSEL *Stirling Castle*, under Captain James Fraser (after whom Europeans named the Island), was shipwrecked and several survivors, including Fraser and his wife, Eliza, put ashore. Subsequent stories about what happened are the subject of much debate, largely due to what many judge to have been mischievous reporting on the part of Eliza, who survived and made money from her tales. Without doubt, the stories resulted in significantly painful consequences for the Indigenous people, who suffered massacres and round-ups by mainland European settlers. The entire island was made an Indigenous reserve in 1860, but following the decision to log its valuable trees, this was revoked in 1863. A sad series of events kept Indigenous people in missions on the island and, subsequently, on the mainland under a dispersal policy, leaving the island free for exploitation.

Logging continued for the next century and removed much of the old-growth forest. Despite an early proposal in 1893 from Australia's peak science body that K'gari be declared a national park, government showed little interest in slowing the intensive logging industry. Even as community groups, both national (Australian Conservation Foundation) and local (Fraser Island Defenders Organisation), increasingly recognised K'gari's natural heritage, a new threat emerged: sand mining for the heavy minerals deposited as part of dune-building processes. The consistent commitment of many individuals—including poet Judith Wright who fought for the protection of Great Barrier Reef and, especially, John Sinclair—raised the bar of protection. The Queensland Government was a significant opponent and conservation groups appealed to the High Court of Australia. When the Committee of Inquiry into the National Estate focused on K'gari as a case study it also heard evidence from Butchulla Elders. The then Australian Government adopted the inquiry's resulting recommendations to prohibit sand mining and recognise K'gari on the newly established Register of the National Estate.

Eventually, all mining leases were relinquished and, by 1991, the damaging logging was finally stopped. The Fitzgerald Inquiry into K'gari was a catalyst for both the cessation of logging and the lodging of a nomination for World Heritage listing—a proposal initiated by the Australian Conservation Foundation and Fraser Island Defenders Organisation in 1984 but resisted by the Queensland Government. Ironically the growing recognition of K'gari as a magnificent natural place after its World Heritage listing created environmental problems. Visitors of all kinds turned up seeking a multiplicity of

OPPOSITE: Dingoes (*wang'ari*) on K'gari have high conservation value and are well protected on the island.

ABOVE: The silently flowing translucent waters of Wanggoolba Creek create a microclimate that sustains subtropical rainforest.

experiences, from wilderness to fishing, from camping to holiday homes, many with conflicting requirements. Management at that stage was dysfunctional, distributed as it was among two local authorities (now amalgamated) and several government departments, all with different aims—and still there was no Indigenous involvement in any capacity. The inquiry had anticipated the need for better management and had recommended a regional management authority. This was also acknowledged as part of the IUCN evaluation. The World Heritage inscription of K'gari in 1992 forced the Australian Government to recognise the need for a formal structure to manage the entire area and for formal community engagement in the management process. A Regional Heritage Advisory Committee was established that included representatives from Aboriginal organisations with an interest in K'gari who gave advice and feedback on management. Community and scientific advisory committees were established over the following eight years, and in 2005 an Indigenous advisory committee provided a formal mechanism for Indigenous input. A culturally appropriate advisory committee was appointed in November 2021 with an independent chair and equal representation of Butchulla people, community representatives and scientific expertise.

The biggest challenge remains managing visitation (up to 500,000 people each year), including the significant environmental impact of large numbers of visitors and the interaction between dingoes and people. Another challenge is ensuring that the fire program is appropriate for the needs of the vegetation communities and the survival of species. Climate change threatens to make such fire management even more challenging and a fire in 2020 burned 85,000 hectares, over half the island. Another issue is biosecurity, with weeds, pests and diseases a major management challenge across K'gari. For example, myrtle rust was first detected on the island in 2013 and is being monitored by the Butchulla Land and Sea Rangers. Investment in staff and other resources has increased since World Heritage listing, as UNESCO's requirement for periodic reporting ensures a clearer focus on desired outcomes and progress towards them. Interpretation has recently expanded with a K'gari Discovery Centre at Kingfisher Bay and the Queensland Government is working with the Butchulla Aboriginal Corporation to develop the island's day-use area, Central Station, for cultural interpretation, and to establish Indigenous-owned and -managed tourism businesses.

The Australian Government intends to nominate an extension to this World Heritage area that will include much of Hervey Bay, the Cooloola sandmass and the Great Sandy Strait between the island and the mainland.

An approaching squall about to wash over Binngih Sandblow near the northern end of K'gari.

TOP: Butchulla Land and Sea Rangers (BLSR) Conway Burns and Myles Broome working with Queensland Parks and Wildlife Service to respond to a fire on K'gari in 2020.

BOTTOM: Matilda, Jodie Rainbow, Blayde Foley, Myles Broome and Chantel van Wamelen (BLSR), with Dr Geoff Pegg and a colleague from Queensland's DAF, setting up a monitoring plot for myrtle rust.

AUSTRALIAN FOSSIL MAMMAL SITES (RIVERSLEIGH/NARACOORTE)

In a stratified rainforest some 22 million years ago, a small shallow pool formed in the limestone environment—clear, calcium-rich water fringed by shrubs and ferns. The forest surrounding the pool was home to an abundance of wildlife, much more so than exists today in similar forests. Possums, thylacines, bandicoots, cuscuses and many others cautiously drank at the pool, some venturing onto a thin limestone crust along its edge. On one occasion, they could hear another animal running fast along the forest floor, with something larger in chase. A small marsupial lion was close to taking its prey, a forest bandicoot. The attempted escape left the bandicoot unbalanced on the edge of the pool and in it went, scrabbling at the collapsing edge. The frustrated lion snorted and left, while the bandicoot swam ineffectively for a while before drowning. Its body sank to the limey mud two metres below and there it remained—the first step in the life of a fossil.

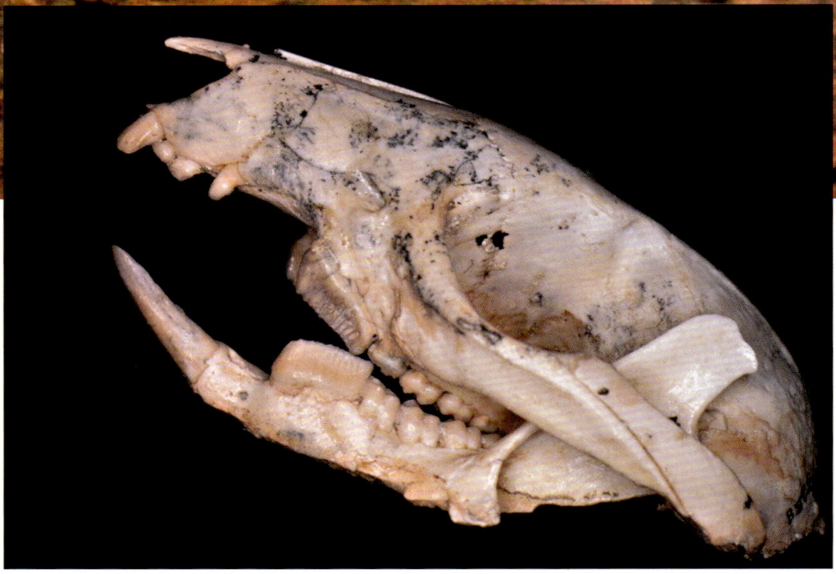

PREVIOUS PAGES: A digital reconstruction of the marsupial *Microleo attenboroughi*, the first fossils of which were discovered at Riversleigh in 2016.

IN THE SEASONALLY DRY north-west of Queensland, some 200 kilometres north of Mount Isa, is that part of the Gregory River catchment known as the Riversleigh fossil site, after the cattle station on which it was discovered. Its 10,000 hectares are on the traditional lands of the Waanyi people, who know the region as part of their *Bujimala* or Rainbow Serpent country. Riversleigh contains one of the world's greatest fossil sites—an astonishing record of Australia's unique mammals over a period of 20 million years. More than 2,000 kilometres south, the Naracoorte cave system in South Australia preserves fossils from the last 500,000 years. These Pleistocene fossils reveal much about the most recent history of Australian mammals, including during the period when humans arrived. The 300-hectare national park features caves beneath naturally occurring eucalypt woodland. The surrounding countryside is ideal for growing wine grapes, which take advantage of the terra rossa soil, derived from limestone. It is limestone that is critical at each site to first form, and then preserve, the relicts of Australia's mammals.

Fossils are formed when the remains of an animal or plant are buried or in some other way preserved—typically in a watery environment—before they are eaten or destroyed. While buried in the mud or silt these remains are subject to the geological process of mineralisation, whereby the organic material is replaced by minerals. Over very long periods, the sediment within which the fossil is buried is subject to many pressures, including compaction, uplifting or folding. Such movements inevitably distort or fracture the fossils and frequently leave only small pieces of bone or impressions. Erosion, often responsible for revealing the fossil at the surface, can itself damage the remains to the point of destruction. In a limestone environment, as at Riversleigh and Naracoorte, animals may accidentally fall into caves or confined pools and be trapped, and therefore die in a good situation for preservation. However, subsequent events rarely allow the preservation of an entire skeleton and it is usually a challenge for palaeontologists to match the various remaining small pieces of bone to others found nearby. More often, all that is left is a fragment, from which scientists must deduce the original species. The varying circumstances of fossil formation may require different investigative techniques to recover the fossil, to date it or to connect it with other fossils.

Riversleigh

Today, the Riversleigh savanna endures long months of dry, dusty conditions. Sparse trees and scattered spinifex are interspersed with rocks of various types and ages. The oldest rocks are the remains of an ancient Precambrian seabed, thrust up and then worn down. Overlaying that are the remnants of marine limestone, perhaps 530 million years old. Most significant are the hills and rocky outcrops of much more recent age, some 15 to 25 million years old from the Oligocene and Miocene epochs. Limestone is abundant here, deposited in dried-up pools and lakes and, in some places, caves. It is within this limestone that the bones of so many creatures have been preserved.

Fossils found at Riversleigh reveal that an astonishing array of mammals lived in the rainforests that once covered the area. They ranged in size from small frogs to huge marsupials the size of a rhinoceros. One genus found here, the *Diprotodon*, was the largest marsupial that ever lived, averaging 3 metres long and 2.6 metres tall at the shoulder. Five or more species of thylacine lived here, ranging from the size of a small cat to a sizeable dog. The Tasmanian tiger was the only thylacine that survived to recent times and it has been extinct since 1936. These were not the only carnivores, as the Riversleigh fauna also included primitive marsupial lions, quite small compared with the better known *Thylacoleo carnifex* and perhaps partway between the *Wakaleo* and *Thylacoleo* branches of these wonderful marsupial carnivores. Experts continue to study the Riversleigh material, which still holds many secrets. Fossils of very strange mammals have emerged from the limestone, but little is known about them beyond the fact of their

OPPOSITE: Riversleigh fossil site from the air.

OPPOSITE, INSET: The skull of a small kangaroo found at Riversleigh.

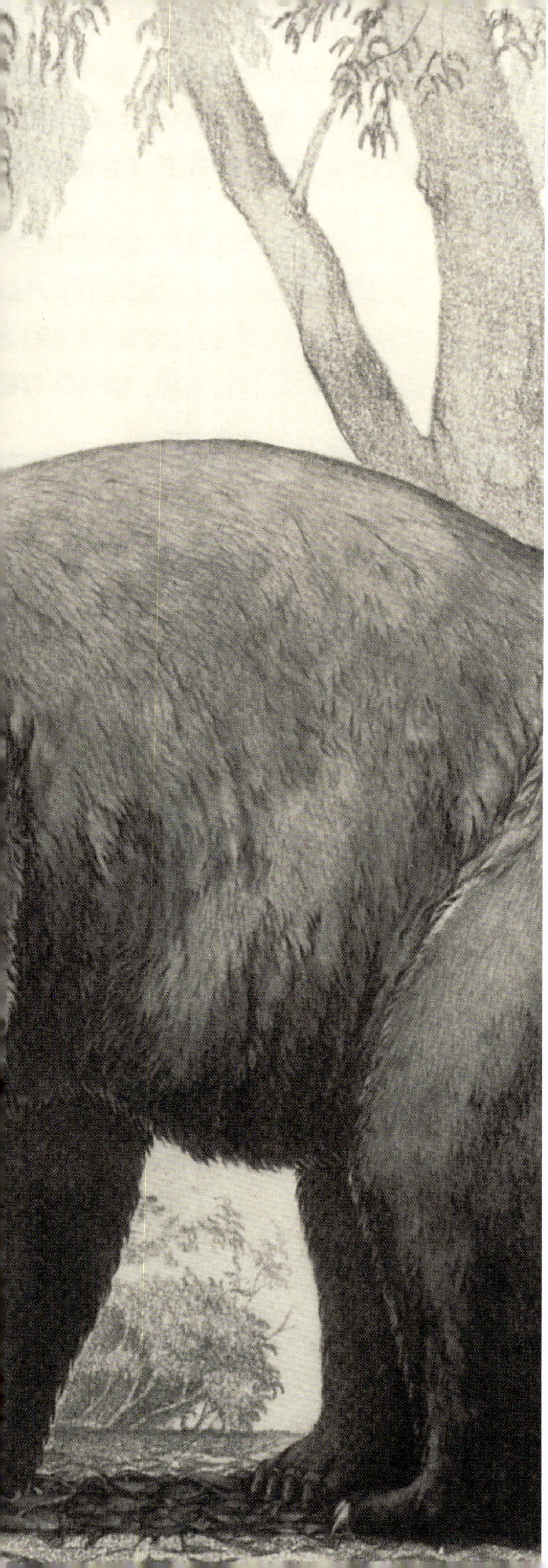

MANY MAMMALS AND BIRDS AMONG THE LIVING FAUNA IN THE WET TROPICS WORLD HERITAGE AREA ARE SIMILAR TO THE FOSSILS FROM THE ANCIENT RAINFORESTS OF RIVERSLEIGH, BUT WITH MUCH GREATER DIVERSITY AT RIVERSLEIGH.

existence. One went by the initial name of 'Thingodonta', as only its strange teeth had been found. Later, when entire skulls were revealed, their strangeness became even more evident and the two Riversleigh species are now classified as an entirely new order of marsupial. Members of the genus *Yalkaparidon*, the only evidence of their existence is at Riversleigh.

Bat fossils at Riversleigh attracted global attention because they represented one of the world's richest deposits covering the Oligocene and Miocene epochs and the more recent Pleistocene. Several species of these micro-bats appear to be closely related to bats found in French fossil deposits. This discovery suggests a novel connection in the Miocene between western European and Australian fauna. It is clear that the habitat through much of this period was rainforest, but that the fauna was vastly richer than the fragments of equivalent rainforest that now remain in Australia. Most species have vanished, including such creatures as flesh-eating kangaroos, numerous herbivores several platypus species and smaller terrestrial carnivores. There is a fascinating link with the living fauna of the Wet Tropics area: a comparison reveals that many mammals and birds among the living fauna in the Wet Tropics are similar to the fossils from the ancient rainforests of Riversleigh, but with much greater diversity at Riversleigh. In the Wet

A digital reconstruction of *Diprotodon*, the largest known marsupial.

I first visited Riversleigh in 1997 when a colleague, Jon Luly, and I were asked to review the outstanding universal value of the site for the Australian Government. I had previously spent time in other parts of Boodjamulla National Park (of which Riversleigh is now a part) and the environment is similar. We were very fortunate that Michael Archer invited us to join his team, which was extracting fossils. We learnt about their field techniques as well as a great deal about the significance of the site. They had developed an extraction technique using detonating cord placed in drilled holes, which ensures precise, limited explosions that crack out blocks of rock without significant damage to the rock itself or the surroundings. The limestone block and other fragments were then removed and sent on the long journey to the laboratory. Using this process, it is sometimes possible to see bone on the surface of fragments, exposed to sunlight for the first time in millions of years and creating excitement at the prospect of the discovery of a new species. I was not surprised that many volunteers had participated in these expeditions.

Tropics, living musky rat-kangaroos, ringtail and brushtail possums, bandicoots and bettongs scamper about the forests; at Riversleigh lie the fossils of their rainforest-loving ancestors. Logrunners and cassowaries illustrate the avian connection between these sites; both feature in fossils at Riversleigh and have living descendants within the Wet Tropics.

Very little of the richness of this outstanding fossil site is apparent when walking across the surface. Rocks are abundant but even an expert eye may overlook fossils. Occasionally some variation in the rock, indicating a fossil, might be visible. Even more rarely, a bone may be revealed by erosion processes, standing out because of its contrasting colour and distinctive shape. Sometimes, excitingly, a cluster of bones and fragments may protrude. Such bones are strongly welded to the surrounding rocks and are extremely challenging to extract. Indeed, initial attempts to remove the Riversleigh fossils by hand proved fruitless and eventually palaeontologists took blocks of fossil-filled stone to laboratories for immersion in weak acetic acid. The acid gradually dissolved the limestone surrounding the fossils, thereby revealing the treasure within. Depending on the make-up of the rock, it generally takes weeks of treatment to free the bones and allow the real work to begin—figuring out the nature of the creature represented by the liberated piece of bone.

This kind of extraction technique means a huge amount of additional work for palaeontologists before they can identify an animal or classify it as an unknown species. They must match the many fragments of bone and, like a three-dimensional jigsaw puzzle, assemble them into a leg or rib, as the case may be. With luck, a complete specimen may be recovered, but often there will be many missing pieces, adding further to the challenge. Gradually, they may uncover fossils of several species and sometimes even evidence about their habitats. Scientists process about 12 tonnes of rock from Riversleigh each year and have revealed over 100

highly diverse species of vertebrate animals. These records demonstrate the wonderful array of life that thrived through the millions of years that Australia drifted north and help us understand how life responded to the drying climate. Michael Archer, Suzanne Hand and Henk Godthelp led the team that first established the significance of the Riversleigh fossils (beginning in 1976) and they continue to reveal ever more wonders.

Fossil site managers face a significant intellectual challenge when they must also maintain the area's heritage value. The nature and extent of fossils at a site are usually only revealed through the extraction of many tonnes of material. Establishing the significance of the site requires a continuing program of extraction to extend understanding and develop the necessary science, which can prove challenging to those charged with protection. In the end scientists must balance the potential for discovering new species with the need to leave an adequate reserve for future activity. It is normal at Riversleigh to leave at least 50 per cent of each excavation site behind, but without extraction the evidence of Australia's mammal evolution and heritage would not be known.

OPPOSITE: Fossil extraction at a site in Riversleigh.

ABOVE: Fossils in limestone at Riversleigh.

Naracoorte

Beneath a series of ancient coastal dunes, along the aptly-named Limestone Coast of South Australia, are the remarkable fossil caves of Naracoorte. Here, fossil species provide many insights into how Australia's mammals responded to climate change. These fossils contain superbly preserved examples of our ice-age megafauna, many of which disappeared over the last 50,000 years. They include the fantastic marsupial lion *Thylacoleo carnifex*, whose ancestral relatives lived at Riversleigh. Unsurprisingly, the site has many well-preserved specimens of modern species, such as the Tasmanian tiger, the Tasmanian devil and various possums, wallabies, mice and bats. The careful study of these fossils has revealed a great deal about changes to Australia's mammalian fauna during the last great ice age and following the arrival of humans.

Despite a century of visitors to these caves, the magnificent repository at Naracoorte remained undiscovered until 1969, when Professor Rod Wells and Grant Gartrell discovered a concealed passage that led to caverns rich in fossils. A palaeontologist at Flinders University, Wells has led research into, and understanding of, this site since his original discovery. In the main chamber, the fossil record covers about 300,000 years. An opening in the ceiling allowed sediment, bones and many animals to fall to the floor of the cavern and begin to accumulate. Gradually, a great cone-shaped pile of material grew, some sliding down the slope and spreading out further over the floor. Many animals died and their bodies remained where they fell to become fossilised. About 15,000 years ago, the cone of debris and bones reached the ceiling and blocked the entrance, sealing the chamber and its amazing contents until its recent discovery. Over the past 50 years, many thousands of specimens have been extracted and these represent at least 93 vertebrate species, including many of the unique Australian megafauna, now extinct.

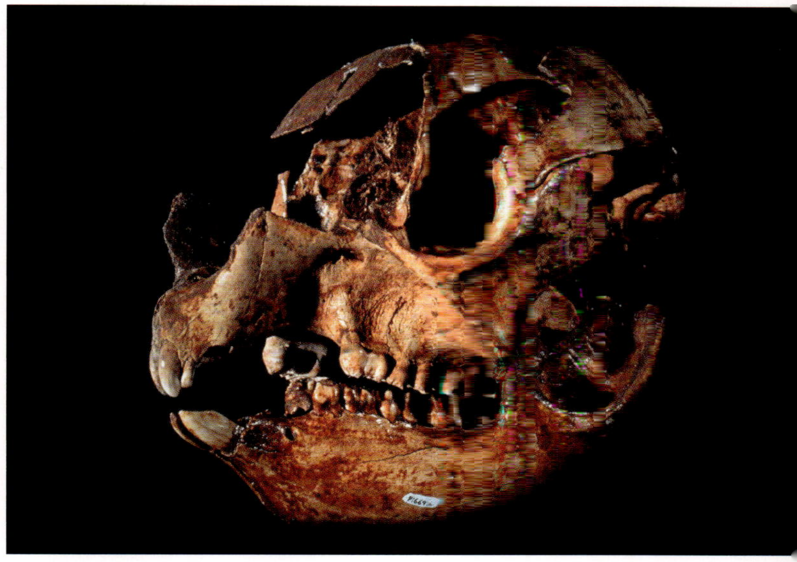

Naracoorte is one of many incredible fossil sites across the world; it is by identifying, protecting and analysing these sites that we learn about the past life of the planet. They help us understand not only the evolution of life on earth, but also the potential effects of changing conditions, from continental drift over eons to climate changes and ice ages over shorter periods.

ABOVE: The skull of a giant leaf-eating kangaroo, found at Naracoorte.

FOLLOWING PAGES: Revealing bone accumulation in the Upper Ossuary, Victoria Fossil Cave, Naracoorte; FOLLOWING PAGES, INSET: The skeleton of a leaf-eating kangaroo, Naracoorte.

OPPOSITE: The entrance to the main hall of the cave at Naracoorte.

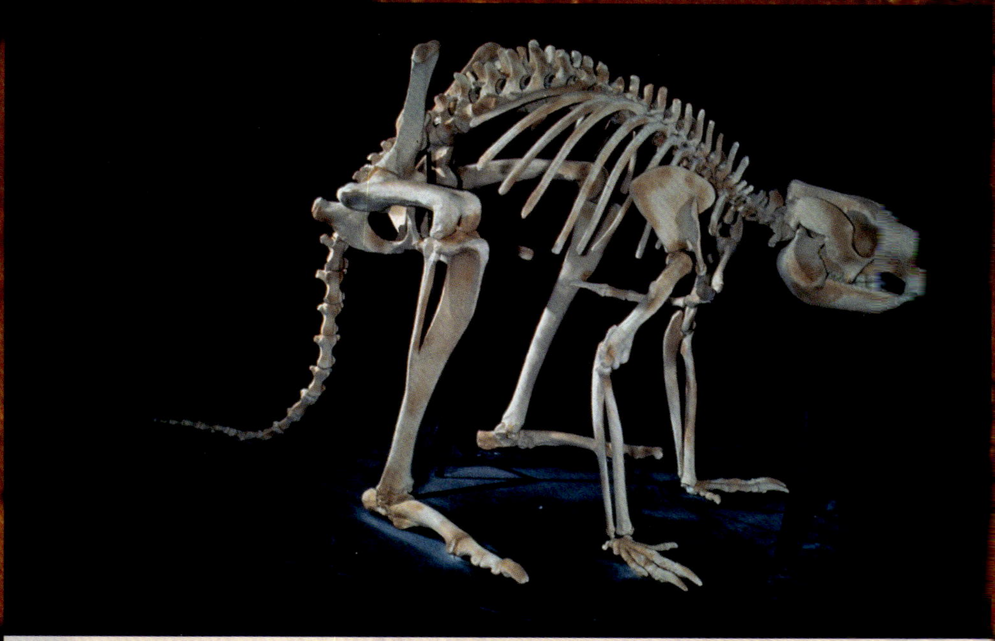

During my 2004 visit to the Naracoorte fossil site to gain further insight into the wonderful World Heritage area, I was fortunate to be shown around by Dr Liz Reed, who has continued much research into the fossils. One of the most impressive elements of my visit was seeing how the site's managers have interpreted the meaning and significance of this site. A tour offered to visitors includes part of the original fossil chamber and places where research continues to unearth the various palaeontological treasures, revealing contemporary excavation techniques. In many places, bones are visible in the exposed strata. There is a magnificent reconstructed skeleton of a marsupial lion, a species that would not have been fun to meet when alive.

Perhaps even more impressive are the reconstructions of these animals on display at the Wonambi Fossil Centre. Here, art and science combine in an interpretation of the natural environment during the Pleistocene, with depictions of the large mammals and other species of the time. To see a giant short-faced kangaroo or the peculiar, tapir-like Diprotodon rendered in fur and form suddenly brings home the astonishing nature of these lost species. The reconstruction of the leopard-sized marsupial lion emphasises its large, pointed incisors and huge thumb claw, perhaps used to disembowel its prey. The extremely large snake, Wonambi naracoortensis, was the last of its family to survive. Growing to an average of six metres long, it was a type of constrictor, with a family so ancient that its ancestors had roamed Gondwana. It had relatives at Riversleigh but the last one vanished, possibly about the time humans arrived. I felt sad that such wonderful animals had become extinct. But, of course, I fully appreciate that extinction is natural and inevitable, as the study of fossils reveals. New life is built on old and all are connected through time.

PATH TO WORLD HERITAGE

THE FIRST REPORT OF A FOSSIL BONE at Riversleigh was in 1900, but very little exploration followed. In part that was due to the remoteness of the site in far north-west Queensland, but it seems likely the site was of less interest than other fossil sites because, for a long time, people thought it was quite young. Then, in 1975, Dr Alan Bartholomai, the director of the Queensland Museum at the time, revealed that he had seen the fossilised jaw of a new species of mammal embedded in a block of limestone but had not been able to extract it for further study. This led Michael Archer, Sue Hand and Henk Godthelp to mount an expedition in 1976 that discovered a rich fossil site with a startling array of bat bones. Although more field trips followed, it was not until 1983 that Michael Archer discovered the major mammal site (known as the Gag site), which by itself yielded evidence of at least 30 mammals never seen before and doubled the previous knowledge of Australia's ancient mammals. After processing in Sydney, this number increased to 58 mammals and a host of other species. In Archer's words, 'this was paradise'.

Following this mega discovery, annual expeditions were scheduled. Journalists and photographers joined many of these trips and so gradually awareness of the site's significance increased. Within a relatively short period, the ongoing productivity and high visibility of the Archer team ensured widespread national attention on the amazing fossils being found. Many television programs, journal articles and documentary films were produced about Riversleigh and its fossils. In 1992, the area was excised from the cattle station to form part of the existing Boodjamulla National Park and be managed as a fossil reserve. The time was ripe to test the prospect of World Heritage recognition and Archer contributed significantly to the development of the nomination.

Naracoorte has a similar story: while the caves and some fossils had been known for a century or more, and visited by many people, it was not until 1969 that Rod Wells first found the exceptional fossils. It was largely through his work and efforts that the significance of the fossils became known and that knowledge conveyed to a wider audience. Eventually, television and other media, as well as scientific publications, helped build public awareness of the fossils and their significance. Professor Wells also contributed significantly to the preparation of the nomination for the Australian Government.

ABOVE: Dr Michael Archer with the partial skull of a species of *Euryzgoma*, an extinct genus of marsupial closely related to the *Diprotodon*.

OPPOSITE: Preparing a rock for explosives that will break it apart at the newly discovered Wholly Dooley site in 2013.

The nomination of these sites proved of great interest to the World Heritage Committee and the IUCN evaluation team. For the first time, a fossil site was proposed that encompassed a combination of separate places spanning a longer window in time than any one area could capture. The Australian Government nomination listed three separate places, including Murgon in south-east Queensland, which has fossils much older than Riversleigh (around 40 million years old). The proposal was to include all three sites as a serial nomination, one in which it would be the combined significance that made the site outstanding. The committee agreed with this proposal but followed the advice of the IUCN that the Murgon fossil site had not been studied sufficiently to demonstrate its significance and that Riversleigh and Naracoorte together were already exceptional. The committee noted that should research confirm the significance of Murgon it could subsequently be added as another part of the serial site.

Although Riversleigh and Naracoorte form a single World Heritage area, they are not managed as such due to the distance that separates them. At Riversleigh, an advisory committee comprising scientists, community members and Waanyi people provides advice directly to the Queensland and Australian government ministers with responsibility for World Heritage. At Naracoorte, there is also an advisory committee. Coordination between Riversleigh and Naracoorte is achieved through collaboration and cooperation between government officials and advisory committees for both locations. There are formal relationships with research institutions and organisations at both sites.

MACQUARIE ISLAND

Once, there was nothing but ocean halfway between Australia and Antarctica, remote from everywhere, the sea floor six kilometres deep. Violently but slowly, the sea floor rose over millions of years, until new land emerged above the waves. It was made of the most unusual of all surface rocks—rocks from the earth's very mantle, the shawl covering the red-hot, flowing innards of the planet. Then the real work of erosion began, shaping and reshaping the rocks, converting some to soil that could receive the seeds of life. Earthquakes reigned supreme, while rain and flowing waters played their part. Eventually, plants from far away colonised the island, but even before that, marine mammals had started to call it home. Birds found it a resting place within the endless seas of the subantarctic region. Soon this became a favoured place for breeding, the island seeming at times to comprise a moving mass of penguins. Macquarie Island was born.

THE ROCKS THAT FORM MACQUARIE ISLAND ORIGINATED IN THE DEEP OCEAN AND WERE SQUEEZED UPWARDS BY THE PRESSURE OF MATERIAL RELEASED BY THE SPREADING SEA FLOOR. THIS JOURNEY, DRIVEN BY TECTONIC UPLIFT AND VOLCANISM, GRADUALLY BUILT TOWARDS THE OCEAN'S SURFACE.

MACQUARIE ISLAND is 1,500 kilometres south-east of Tasmania and about halfway between Australia and Antarctica. Located in the fierce, constant winds of the Furious Fifties, it is 34 kilometres long and between three and five kilometres wide, with an area of 12,875 hectares. The total area of the World Heritage site is 540,000 hectares, mostly made up of the marine reserve that protects the adjacent waters of the Southern Ocean. While a dominant reason for the island's inscription on the World Heritage List is its exceptional geological properties, it is also recognised for the spectacular wildlife that uses the island for breeding. These inhabitants include many species of bird and mammal—collectively seabirds may number 3.5 million, one of the greatest concentrations in the world. It is the only breeding location for the endemic royal penguin—the largest colony has 230,000 pairs, occupying a vast area of the coastal shelf at Hurd Point, the southernmost tip of the island. The overall population estimate is 750,000 nesting pairs (in 2016). This number is lower than the census in 1984 but measurement technique differences may account for that.

The rocks that form Macquarie Island originated in the deep ocean, some 6,000 metres below sea level, and were squeezed upwards by the pressure of material released by the spreading sea floor. This journey, driven by tectonic uplift and volcanism, gradually built towards the ocean's surface. The island tops a very long underwater ridge with exceptionally steep slopes that plunge abruptly to the sea floor a very short distance offshore. Geologists recognise one particular sequence of rock as an ophiolite. This is a section of igneous rock from the earth's crust and upper mantle that has pushed to the surface, accumulating layers of basalt and marine sediment on its journey. Ophiolites provide crucial evidence of both sea floor spread and continental drift. In most cases, ophiolites are found near the continental rocks over which they have moved. In the unique example of Macquarie Island, there is no associated continental rock, so the rare mantle rocks from the ocean floor are ideal for analysis.

PREVIOUS PAGES: Royal penguin rookery, 1984.

Although the spreading sea floor ridge began to grow many millions of years ago, the deep mantle rocks did not reach the sea surface until 600,000 years ago. Here, the rocks were subject to further erosion, but, unlike on other subantarctic islands, there were no glaciers to erode or volcanoes to cover up the mantle rock. Uplift has continued and produced three peaks over 400 metres above the sea. Erosion has shaped both a coastal terrace and steep escarpments that inhibit access to the central undulating plateau. Macquarie Island vegetation is low—mainly a few extremely hardy herbs and some tough tussock grasses. It is subject to the prevailing winds, bleak and powerful westerlies that rarely subside. Relentless pounding by the Southern Ocean has produced many rugged bays, sea stacks and rocky reefs. The plateau itself is scattered with lakes and pools; at its edges, streams drain down to the coastal platform, which is rich in boggy features. The annual rainfall (900 millimetres) is spread throughout the year, keeping conditions moist, and cold temperatures can mean snow at any time. By most standards, this is a fierce environment in

OPPOSITE: Boulders near the summit of Mount Hamilton are surrounded by the only kind of vegetation on the windy island: low and extremely hardy.

Elephant seals bask on the island's shores during breeding season (September–November).

which to live and yet life is the second compelling reason for this remote island's World Heritage inscription.

In commenting on the island's biological significance, the World Heritage Committee noted that 'Macquarie Island provides an outstanding spectacle of wild, natural beauty with huge congregations of penguins and seals'. Most of the wildlife on the island clusters in places along the shoreline or on the terrace. Most obvious are the huge elephant seals that come ashore to breed in spring. Each dominant male, weighing up to 3,000 kilograms, claims a section of beach as his own and defends it fiercely. Within the territory are basking females, the focus of the beachmaster's interests. From Macquarie Island, they head south to Antarctic waters to feed before returning in late summer to moult. They are named for their large snout, used to make their impressive roars during the breeding season. Macquarie Island hosts about 70,000 of these astonishing mammals during the breeding season, as well as large numbers of fur seals (three species) and occasional leopard seals. Offshore killer whales visit regularly, hunting for young seals.

Birdlife on Macquarie Island is similarly spectacular. Along the shoreline, several species of penguin congregate in

Wandering albatross courting.

enormous numbers on beaches and in bays. Lusitania Bay, on the sheltered eastern side of the island, houses a colony of the large king penguins. Over 250,000 birds breed here between November and May. The chicks stay on the island through winter until the moult is finished and their parents return occasionally to feed them after foraging for long periods. There are always some king penguins on the island. Over the summer months, about 37,500 pairs of southern rockhopper penguins breed on the island, a small proportion of the global population. Royal penguins only breed at Macquarie Island and, at 750,000 pairs, are the most abundant penguin species on the island. They seem to prefer breeding in large colonies, which are alive with sound and movement all day. Parents forage in the deep waters south-east of the island, returning to feed the chicks every day or two. Following the post-breeding moult, royal penguins leave Macquarie Island, but where they spend their time between leaving the island and returning for the next season's breeding remains largely a mystery.

Apart from penguins, there are many other species of seabird on Macquarie Island. Four species of albatross breed each year on the steep cliffs: the light-mantled sooty albatross (at 2,000 pairs, the most numerous), black-browed albatross, grey-headed albatross and wandering albatross. The giant and white-headed petrel, endemic Macquarie Island cormorant, Antarctic prion, Dominican gull and giant skua are all present for much of the year, the last feeding on penguin eggs and chicks. A number of other birds visit occasionally, including another two species of penguin. Sadly, human activity has caused the extinction of two endemic birds: the Macquarie Island parakeet (closely related to the Lord Howe Island parakeet, also extinct) and the Macquarie Island rail (related to the Australian buff-banded rail).

While the geological attributes that make Macquarie Island so special remain important, for most people it is the spectacular wildlife that is cause for celebration. Few places provide such spectacles of wildlife being wild. Macquarie Island is naturally remote from immediate disturbance from humans, although growing tourism activity will require careful management to ensure the continued presence and stability of wildlife. The albatross that provide such a spectacular sight as they glide into their nests are much safer on Macquarie Island now than in the past, though they are still at great risk from the long-line fishing that has decimated millions across the globe.

King penguins near Sandy Bay.

My first connection with this World Heritage area was while preparing a lecture on resource management practices at the beginning of European settlement. I was astonished to read of the destruction of wildlife by desperate Australian sealers in the first few years after the discovery of Macquarie Island in 1810. It seemed shocking that such a place, never previously visited by humans, could suddenly be so ravaged. It is a black spot in our environmental history and sadly all too common for remote islands. Sealing was not just a brutal and brutalising business for those engaged in it, there were extraordinary consequences for the other island inhabitants. Rabbits, introduced by sealers as a food source, devastated the island vegetation, while cats and rats drove two bird species to rapid extinction. The recent efforts to remove introduced pests, while unable to reverse extinctions, are a positive story of human engagement with Macquarie Island. It is now rabbit-, cat- and rat-free. It has to stay so.

Dean Miller, a marine biologist and former PhD student of mine, made a splendid documentary about the year he spent on Macquarie Island studying the recovering populations of fur seals. He also filmed a gruelling expedition he made to Hurd Point to see the breeding royal penguins, now well recovered from their previous decimation. His images were wonderful and the title of his film, Edge of Nowhere, seems very appropriate. One compelling image from the video that stays in my memory is of Dean sitting on a beach, surrounded by a mob of king penguins. They look curious as they slowly parade around him, standing head height. It is a brilliant representation of wild creatures connecting with a human.

PATH TO WORLD HERITAGE

THE FIRST RECORDED SIGHTING of the island was in 1810. Having destroyed the Australian, Tasmanian and New Zealand populations of fur seals, sealers were seeking new stocks. Within 18 months, over 120,000 fur seals had been killed and numerous sealing gangs were stationed on the island. By 1815, the plunder was complete and that year only 6,000 fur seals were found. Attention turned to elephant seals, which were killed and their blubber reduced to oil on the island. Some sealers turned to penguins and first king penguins then royals became targets of the massive slaughter of wildlife. Douglas Mawson, who stopped at Macquarie Island in 1911, was so concerned that he called for protection of the island and its remaining wildlife. Eventually, the exploitation ceased in 1920.

Since March 1948, Australian National Antarctic Research Expeditions (ANARE) has operated a scientific research station on Macquarie Island, providing a long-running investment in better understanding the island's wildlife. The research undertaken here has provided an excellent foundation for the island's long-term conservation, providing a model for conservation activities around the world. The island became a state reserve under the Tasmanian Government in 1972, further improving its conservation status. UNESCO named the island as a biosphere reserve in 1977 and in 1980 it was listed on the Australian Register of the National Estate (now National Heritage).

OPPOSITE: Mount Kidson.

ABOVE: Reminders of Macquarie Island's seal-hunting past.

Australia nominated Macquarie Island for consideration by the World Heritage Committee in 1992, but only on geological grounds. The IUCN recommended that Australia consider the property as an oceanic island ecosystem representative of the subantarctic biogeographic realm. The World Heritage Committee endorsed this view and suggested that Australia consider working with New Zealand for a joint nomination of subantarctic islands. It decided not to inscribe the property at that time.

Australia persisted with a sole nomination and retained its focus on geological rather than biological criteria. A new nomination along these lines was put forward in 1996. At that time, the IUCN again recommended that the decision remain not to list until more attention was given to the biological significance of Macquarie Island. The Australian Government's reluctance to conform with this request was not explained. Even so, the final nomination did include reference to biological attributes and this inclusion led the committee to inscribe the island under both geological and biological criteria.

Although geological significance was a strong factor in the island's nomination, research and management plans have centred almost entirely on biological elements. Management agencies faced huge tasks in repairing the severely eroded and damaged environment and removing feral pests. They deserve great credit for their effort and the success of the recovery programs, which resulted in the wholesale eradication of cats by 2000 and of rabbits, rats and mice by 2011. This is a remarkable achievement given the challenging natural environment. Agencies continue to monitor the island, documenting its recovery and guarding against the reintroduction of pests.

ABOVE: Douglas Mawson beside a whale boat at Caroline Cove.

OPPOSITE, BOTTOM: A fur seal is captured and harnessed to be weighed and tagged at a colony at Secluded Bay, 1998.

OPPOSITE, TOP: Examining a royal penguin in 1948, when the Australian National Antarctic Research Expedition established Macquarie Island station.

FOLLOWING PAGES: A colony of king penguins huddle on a shore of Macquarie Island.

HEARD AND McDONALD ISLANDS

With its howling gales, the Furious Fifties challenged the sailing skills of every nineteenth-century captain. The temperatures are extreme even in summer and the wind chill deadly. Constant westerlies cause huge swells year-round. From the early 1800s, seal hunters travelled 450 kilometres from Desolation Island to the bleak and dangerous Heard Island, putting a team ashore for the summer while the sailors returned to the relative safety of Desolation. Left in Spit Bay with almost no protection from the elements, the hunters had a narrow focus: kill elephant seals, cut them up, boil the blubber into oil, pour the oil into barrels and keep going all day, every day. These were extraordinary men, tough and strong, survivors perhaps. One might admire their persistence, their indomitable spirit, their bravery, even while being horrified by the brutal trade they practised. This was the start of recorded human interaction with the remote Heard and McDonald islands.

RISING FROM THE DEPTHS of the remote subantarctic waters, Heard Island is a small protrusion of land located 4,100 kilometres south-west of Perth and only 1,500 kilometres north of Antarctica. It is one of the wildest and most remote places on earth, deep in the Furious Fifties and subject to consistently harsh weather. Heard Island (36,800 hectares) is dominated by Big Ben. An active volcano, covered by glaciers, at 2,745 metres Mawson Peak is Australia's highest mountain outside of Australian Antarctic Territory. To the west, the much smaller McDonald Island has steep cliffs rising to about 230 metres and in recent years volcanic activity has extended its area to about 2,500 hectares, uniting it with the formerly separate Flat Island. The profile of Heard Island reflects the bleak environment, with a narrow fringe of low and extremely tough vegetation along the shore, bare slopes rising starkly and steeply, some dotted with penguin colonies, and vast areas of snow-covered slopes and glaciers soaring to Mawson Peak.

As might be expected, most of the plants on the island are low-profile mosses, liverworts and lichens but there are 12 vascular plants as well. To survive in such a harsh

environment, plants must be extremely hardy. Simply getting to these remote islands is an extraordinary feat and as such only 12 species of flowering plant are known to grow on Heard Island. These are all low-growing herbs or grasses and none are endemic to the island. They occupy the small, ice-free areas at lower elevations near the coast. Most have adapted to the environment by growing smaller leaves, but one species, the Kerguelen cabbage, has large, thick, hairy leaves and produces large, flowering clusters. It is a relative of the common cabbage, and early European sailors ate the leaves to protect against scurvy. It is now known that the leaves contain an essential oil rich in vitamin C. Most of its close relatives are pollinated by insects but this species is wind-pollinated, thought to be a response to living in an area largely free of insect pollinators.

Australia's most remote World Heritage site covers a vast area of the Southern Ocean (658,903 hectares), of which only a small part is land. Heard Island has an incredibly rich wildlife presence, including seals and penguins and other seabirds. After Europeans first came across it, in 1853, many

PREVIOUS PAGES: Morning light on fresh ice at Heard Island.

ABOVE: Big Ben, a volcanic massif, containing the highest Australian mountain peak outside the Australian Antarctic Territory.

sealing vessels and gangs, mostly American, visited Heard Island to take skins and, more significantly, extract elephant seal oil. Over a 30-year period, more than 100,000 barrels of oil were produced and the elephant seal population almost exterminated, thereby destroying the industry. The remnants of these sealing days are scattered on the narrow coastal fringe, the only place even remotely suited to human settlement and the location of the vast concentrations of seals and other wildlife that have now recovered from the past slaughter.

The 1982 IUCN review of the world's greatest natural areas identified Heard and McDonald islands as potential World Heritage areas and in particular drew attention to their wildlife. The ecological wellbeing of the islands' animal and plant life is largely due to the relative lack of introduced pest species, which have been so problematic on other remote islands. In addition, no extractive industry is allowed and there is now no permanent settlement, with only rare and minor temporary visits by humans. This provides an exceptional opportunity for the maintenance of biological processes and the continuing assessment of conservation outcomes away from the direct disturbance of people. This World Heritage site is also an outstanding example of ongoing geological processes, particularly of active volcanoes—Heard and McDonald islands have the only active subantarctic volcanoes in the world. Not only that, the glaciers that have formed on Heard Island are relatively shallow and fast flowing and respond quickly to climate change. This makes them important sites for monitoring changes to our warming planet as they occur.

OPPOSITE: Large numbers of Macaroni penguins use Heard Island for breeding in summer months.

ABOVE: Lava flow on Mawson Peak, Heard Island, 2017.

Like other enthusiastic birders, I have from time to time participated in a pelagic birdwatching trip. This involves heading off in a boat—from one of many appropriate harbours around southern Australia—to a point near where the continental shelf drops off into much deeper waters. Here, the rich availability of food attracts many ocean-roaming birds and enables birders to observe and photograph species that rarely come anywhere near the coast. In February 2015, I went with others on a trip from Albany in the far south of Western Australia. In the course of eight hours we travelled well off the coast to where the water depth increased dramatically and ocean upwelling brought food to the surface, and we enjoyed the sight of many seabirds.

One of the most exciting birds for me was a wandering albatross, my first sighting of this magnificent ocean wanderer. It is possible that this bird has visited Heard and McDonald islands, but this was the closest I got to the World Heritage area. We also enjoyed seeing storm petrels, shearwaters and other species of albatross. Later that year, I saw more wandering albatross off Kiama in New South Wales and then in February 2016 was delighted to see several handsome black-browed albatross on a trip off Eaglehawk Neck in Tasmania. I wonder whether these black-browed albatross have visited Heard Island, where there is a colony of 700 breeding pairs.

Heard Island is one of the world's least anthropogenically disturbed locations. A number of bird species live on the islands, especially during breeding seasons. Macaroni penguins, one of the crested species, are abundant on Heard Island, with significant summer breeding colonies. A close relative, the rockhopper penguin, also lives on Heard Island but its numbers are lower than in the past and seem to have been reducing further in recent times. Given the remoteness and rugged nature of this site, very limited research has been done on any of the wildlife. The larger king penguins are present, with one colony at Doppler Hill made up of 25,000 breeding pairs. The gentoo penguin also breeds at Heard Island, but this globally widespread species is found here only in small numbers. Another two species of penguin have also been recorded on the island, but there is no evidence of breeding to date. Among the flighted seabirds, some 15 species are known to breed on Heard Island and another 28 species have been recorded visiting or living there. The petrels dominate, with 7 breeding species and another 15 or so species recorded in the area. The wandering albatross— the world's largest, with a three-metre wingspan—has been recorded breeding on the island, but numbers are low. The much smaller and more abundant black-browed albatross also breeds on the island. As with most remote subantarctic islands, the coastal shelf provides extensive breeding areas for the seabirds.

Surprisingly, despite the harsh conditions, 127 species of invertebrates are known to inhabit Heard Island, including several endemic beetles, flies, a weevil and a springtail. Remoteness and glaciation has kept insect diversity low, with just 33 species, mainly parasites on birds and seals. There are 12 free-living flies and beetles, which are found in a variety of habitats where plants occur. The beetle population shows unique evolutionary adaptations to the environment. Weevils, for example, have a high resistance to temperature extremes and their body sizes diminish with altitude. Severe winter conditions are associated with

ALTHOUGH THERE WILL NEVER BE MANY VISITORS TO THIS WORLD HERITAGE AREA, ITS RELATIVELY PRISTINE CONDITION AND ABUNDANT WILDLIFE, ITS SPECTACULAR GEOMORPHOLOGY AND ONGOING GEOLOGICAL PROCESSES, AND THE BLEAK BEAUTY OF THE PLACE MAKE IT AN OUTSTANDING ADDITION TO THE AUSTRALIAN SITES.

a dramatic drop in numbers of insects, to less than five per cent of the summer population. The largely intact and pristine ecosystem makes the island ideal as a natural laboratory for examining environmental change as a result of regional climate warming. Seven species of seal have been identified on Heard and McDonald islands, with breeding colonies of three species. The non-breeding winter population of leopard seals (in excess of 1,000) is the highest on any subantarctic island. Elephant seals are even more abundant, and live mainly on the more sheltered eastern side of the island. Both Antarctic and subantarctic fur seals breed on the islands, the latter an apparently recent arrival building in numbers. There is scope for significantly more research on the wildlife of this remote area. Although there will never be many visitors to this World Heritage area, its relatively pristine condition and abundant wildlife, its spectacular geomorphology and ongoing geological processes, and the bleak beauty of the place make it an outstanding addition to the Australian sites.

PREVIOUS PAGES: Elephant seal pups in a secluded corner of Atlas Cove, Heard Island.

OPPOSITE: The black-browed albatross is one of 34 bird species recorded on Heard and McDonald islands.

PATH TO WORLD HERITAGE

ORIGINALLY CLAIMED BY THE BRITISH, these islands passed to Australia as a territory in 1947 and have since been managed by the Australian Antarctic Division, which is based in Hobart. Although initially the focus was on exploiting the wildlife—especially seals and penguins for oil—attention later turned to science and exploration. In 1929, Sir Douglas Mawson visited the island on his way to Antarctica as leader of the British, Australian and New Zealand Antarctic Expedition. When control of the islands transferred to Australia, the Australian National Antarctic Research Expeditions set up a station at Atlas Cove and mounted a two-year settlement and research project. There have since been many research expeditions, including into potential fisheries.

ABOVE: Eight scientists sheltered in this hut on Heard Island in November 1929 during a prolonged blizzard.

OPPOSITE: Making some noise in front of an Australian weather station in 1948.

The Australian Government nominated Heard and McDonald islands for World Heritage inscription in 1991. The IUCN evaluation at the time recommended against listing until further information was provided about legal protection of the site, marine conservation values, progress on a management plan and the area's boundary. At least partly in response to this, a management plan was prepared that came into effect in 1996. Subsequently, Australia prepared a new nomination that allayed UNESCO's previous concerns and the site was inscribed under two criteria.

Following the listing, the Heard and McDonald Islands Marine Reserve was established in 2002. It later expanded in 2004, acknowledging the high conservation value of the surrounding waters. A treaty with France has been in place since 2003 to provide cooperative surveillance of Heard and McDonald islands and the Kerguelen Islands, 450 kilometres to the north-west. The focus of management by the Australian Antarctic Division is on protecting the nationally and internationally important natural values and heritage places.

OPPOSITE: Botanists Jenny Winham (left) and Patricia Selkirk collect plant samples on Heard Island in 2000.

ABOVE: A light amphibious cargo vessel retrieves gear from Spit Bay Camp at Heard Island, with Stephenson Glacier in the background.

GREATER BLUE MOUNTAINS AREA

Throughout the twentieth century, mateship and the bush were two classic Australian ideas. Celebrated in fiction (Henry Lawson), poetry (A.B. 'Banjo' Paterson and Dorothea Mackellar) and art (Albert Namatjira and Tom Roberts), they reflected a growing national interest in the Australian experience. The two went together no better than in the life of Australia's earliest conservationist, Myles Dunphy. He epitomised the tradition of healthy outdoor living based on the inexpensive recreation of hiking and camping. Beginning in 1912, he hiked and camped the relatively unexplored Blue Mountains, developing ideas about conservation and even technology for camping. With mates, he set up the Mountain Trails Club and later the Sydney Bushwalking Club and these in turn led to a broader knowledge of the unique Australian wilderness and a commitment to conservation among the general population. Dunphy shared his passions and wrote about his ideas. He formulated and promulgated the idea of a Greater Blue Mountains National Park and celebrated the notion of protecting the bush for future generations. There is nobody more responsible for this national treasure, though it took the contributions of many, including the Colong Foundation and other community organisations, as well as three decades of effort, to finally achieve Dunphy's dream.

THE GEOMORPHOLOGY IS REMARKABLE: SLOTTED CANYONS, CLIFFS, WATERFALLS AND OUTCROPS REFLECT THE LONG PROCESS OF LANDFORM EVOLUTION.

THE GREATER BLUE MOUNTAINS AREA World Heritage property was inscribed on the World Heritage List in 2000 for its natural heritage, especially its biodiversity. It comprises eight reserves: Blue Mountains, Gardens of Stone, Kanangra-Boyd, Nattai, Thirlmere Lakes, Wollemi and Yengo national parks, and Jenolan Karst Conservation Reserve. Covering more than one million hectares, the Greater Blue Mountains Area has an astonishing array of intact habitats and species, all the more remarkable for being so close to Australia's largest city. Over 65 per cent of the area is wilderness; the crenulated and deeply eroded edges of ochre-coloured sandstone plateaux descend sharply into deep valleys and rivers. The geomorphology is remarkable: slotted canyons, cliffs, waterfalls and outcrops reflect the long process of landform evolution, including areas of limestone with spectacular cave formations.

Diverse geology has provided opportunities for plants to occupy a great range of habitats and in so doing develop a wonderful array of different species. In particular, plants in the family Myrtaceae, including eucalypts, have thrived and diversified, and demonstrate how flora changed as Gondwana broke up and Australia drifted north. The eastern escarpment connects this site with the Gondwana Rainforests of Australia World Heritage site, but the plant story in the Greater Blue Mountains is about eucalypts. As the Australian continent dried out during its long journey north, rainforest species could not survive the increasingly harsh conditions and retreated to the few small pockets of habitat that remain today. By contrast, these were perfect climatic conditions for the sclerophyll species, which now dominate the entire continent. Sclerophylls have tough, leathery evergreen leaves—adaptations to better survive seasonal drought. Such plants grow extensively in open forests and woodlands across Australia, with most endemic eucalypts and wattles displaying this property. Another similar adaptation, scleromorphy (small, rigid and tough leaves), helps plants thrive in low-nutrient soils. Much of the dominant flora of the Greater Blue Mountains has these adaptations, developed over millions of years. There are over 100 species of eucalypt in this relatively small area of Australia, including several forest types, woodlands and some stunted mallee forms, all showing distinctive adaptations to climate change.

In 1836, Charles Darwin visited the Blue Mountains towards the end of his voyage on the *Beagle*. He was an acute observer and made specific descriptions about the eucalypt-dominated bush he saw in Australia:

> *The trees nearly all belong to one family; and mostly have the surface of their leaves placed in a vertical, instead of as in Europe, in a nearly horizontal position: the foliage is scanty and of a peculiar pale green tint, without any gloss. Hence the weeds appear light and shadowless ... the leaves are not shed periodically ... the bark of some [trees] falls annually, or hangs dead in long shreds, which swing about with the wind, and give to the woods a desolate and untidy appearance.*

The Greater Blue Mountains World Heritage property offers an excellent opportunity to share the story of eucalypts—Australia's most significant plant—with the rest of the world. Eucalypts have many characteristics that make them globally unusual, as was noted by Darwin. They are a distinctly Australian flora but have been widely planted elsewhere in the world. The mature leaves of most eucalypts hang vertically, allowing a high percentage of sunlight to penetrate the canopy and reach the ground, supporting a rich understorey of plants.

PREVIOUS PAGES: Looking out to the Three Sisters with Jamison Valley beyond.

OPPOSITE: Wentworth Falls drops 100 metres down the sandstone of the Blue Mountains plateau.

Eucalypts have also adaptated to survive fire. They are able to sprout at the base and can produce shoots from special epicormic buds on the trunks and branches following fire. Thick bark on many species acts as protection, while woody capsules protect seeds during fires and delay opening so that seeds have a better chance of survival. Somewhat surprisingly, they also have some qualities that promote fire, such as a high content of volatile oils in their leaves, leaf litter that breaks down very slowly and an open canopy. Eucalypts also have an exceptional capacity to thrive in low-nutrient conditions, forming forests of high biomass despite Australia's poor soils. Eucalypt forests seldom contain other types of trees, but do usually have several different species of eucalypt making up the community (in contrast to many single-species-dominant forests elsewhere). They cover a remarkable range of environments from sea level to the subalpine tree line, from the tropics to the coolest temperate areas. Almost all of the 900 species occur only in Australia. Most of the Greater Blue Mountains World Heritage property is dominated by eucalypts that, because of their isolation from human activity, are in excellent natural condition.

This isolation also protects many other plants and animals. One particularly famous example is the Wollemi pine, sometimes described as a living fossil because it was present during the age of the dinosaurs some 100 million years ago. Until its discovery (by Europeans; it would have been familiar to the area's Indigenous groups) in remote, virtually inaccessible gorges in the Greater Blue Mountains, it had only been known as a fossil. This relict population—of perhaps the rarest, if not the most significant, plant in the world—was recorded by David Noble while he was bushwalking in 1994. There are fewer than 100 mature examples of this handsome tall conifer,

OPPOSITE: Sclerophyllous eucalypts dominate the Greater Blue Mountains World Heritage Area.

ABOVE: With highly flammable oils in their leaves, eucalypts promote fire even while they are able to survive it.

with its unusual dark green foliage and distinctive bark, remaining in Australia (and therefore the world). It is related to the *Araucaria* genus (which includes hoop, Norfolk Island and bunya pines), which also has species growing in South America, providing yet another link across the great Gondwana continent.

Sixty-nine species of native mammal have now been recorded within the area, 28 of which are threatened, including the spotted-tailed quoll, yellow-bellied and squirrel gliders, and the brush-tailed rock wallaby. Nine species of bat are listed as threatened, as are 15 species of bird. Over 250 species of bird have been recorded, including the superb lyrebird. Twenty-nine of Australia's 77 species of honeyeater have been recorded in the World Heritage area, most of which feed on the blossoms of eucalypts and other plants. One of the 74 species of reptile is found only in this area, the Blue Mountains swamp skink. There are 36 frog species known to live in the region, but several are threatened and perhaps one is extinct (the green and golden bell frog). The invertebrate fauna is not well known, but includes at least 120 species of butterfly and an estimated 4,000 moth species. The area is also home to a rich and specialised cave fauna.

The erosion of the sandstone plateau has produced some exceptional natural sculptures, including the famous Three Sisters (part of the Blue Mountains National Park). But there are many other outstanding features and vistas resulting from the long natural transformation of the country. Within the Gardens of Stone National Park are the unusual pagoda rock formations, a product of natural weathering. The sandstone layers that form the plateau include different types of rock and some are much more erodible than other, more resistant layers. One consequence of this is differential weathering, whereby the softer material is cut out more deeply beneath a resistant layer. Sometimes in this process the forms left may be quite rounded, as in the beehive formations that dominate Purnululu World Heritage Area in Western Australia. The Blue Mountains forms—although

The spiky Wollemi pine is sometimes described as a living fossil because it has existed for some 100 million years.

somewhat similar in their banding appearance—are blocky, with angular edges to the undercutting, developing at times a distinctly pagoda-like appearance.

The exceptional biodiversity and geodiversity attributes of the property are complemented by numerous others, including water systems, wilderness, recreation, natural beauty and Indigenous cultural heritage. The area has significant cultural value, having been occupied by Indigenous people for many thousands of years. The Greater Blue Mountains Area spans the Country of the Darkinjung, Dharawal, Dharug, Gundungurra, Wonnarua and Wiradjuri peoples. Indigenous people, through ongoing practices that reflect both traditional and contemporary presence, continue to have a custodial relationship with the area. Rock art, including hand stencils, can be found in many of the sandstone environments. There are also impressive rock engravings showing kangaroos and other animals. Deep in the Wollemi wilderness, there are many sites rich with Indigenous rock art, including the outstanding Eagle's Reach cave. In this remote area, the art is well preserved and remains within its original bushland setting. It has been suggested that this area has long been a special meeting place for Indigenous people from different language groups.

Australian history emphasises the challenge faced by the invading Europeans in heading west from Sydney. The Blue Mountains bottled up the colony on the coastal plains. Following rivers inland inevitably led to narrowed valleys and vertical cliffs with little chance of egress. Very limited opportunities existed for a passage over the mountains and so there remains, even today, only one major communication link through the mountains for road and rail. Settlement is largely confined to that narrow link, with wilderness to the north and south. Geomorphology and geography have conspired to overcome the pressures from the sprawling metropolis and to preserve an astonishing wilderness.

ABOVE: Rock art in Red Hands Cave is thought to be between 500 and 1,600 years old.

OPPOSITE: The Zig-Zag Railway through the Blue Mountains was built to connect Sydney to Bathurst and other western New South Wales towns.

FOLLOWING PAGES: The view at Wentworth falls with Mount Solitary in the distance.

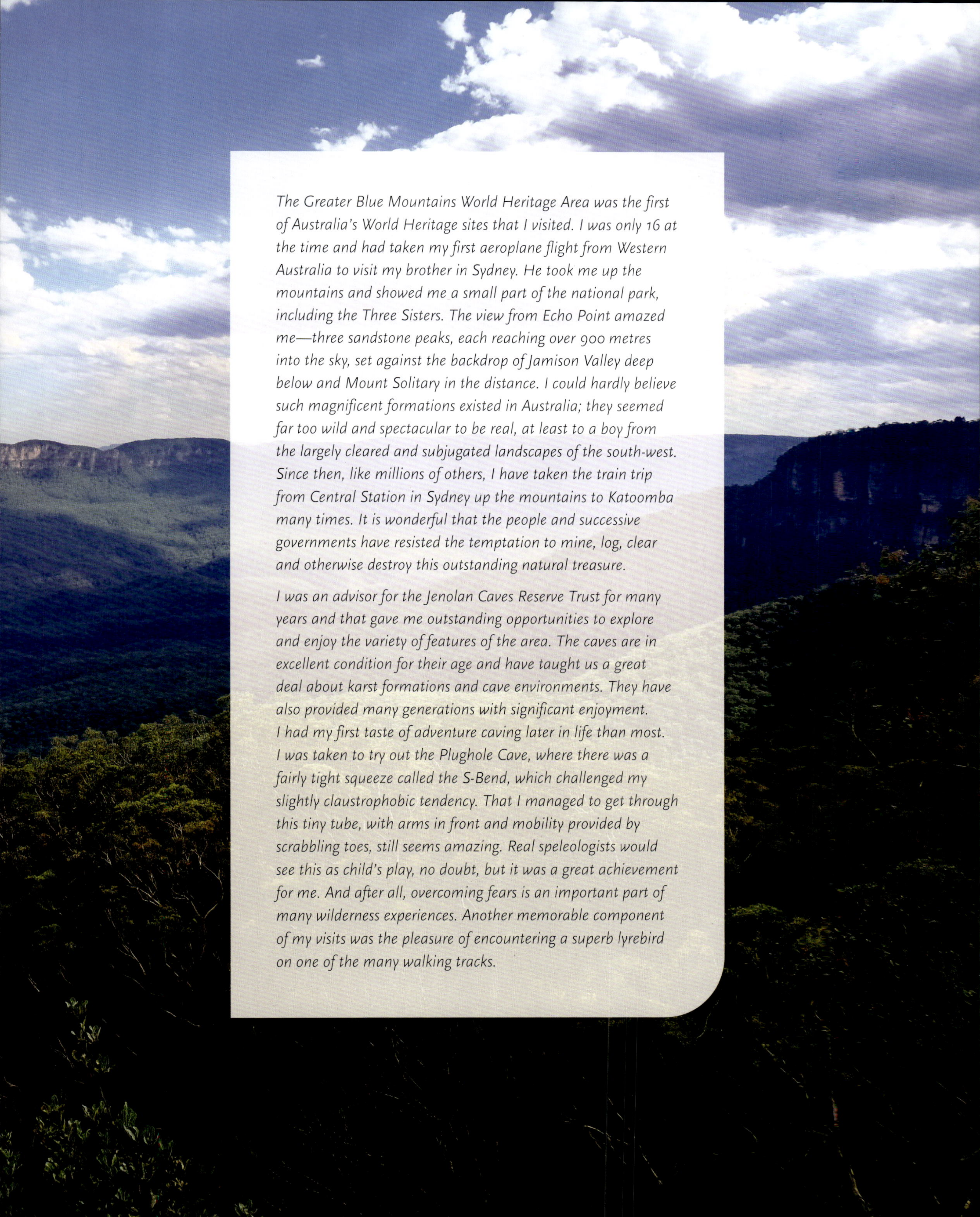

The Greater Blue Mountains World Heritage Area was the first of Australia's World Heritage sites that I visited. I was only 16 at the time and had taken my first aeroplane flight from Western Australia to visit my brother in Sydney. He took me up the mountains and showed me a small part of the national park, including the Three Sisters. The view from Echo Point amazed me—three sandstone peaks, each reaching over 900 metres into the sky, set against the backdrop of Jamison Valley deep below and Mount Solitary in the distance. I could hardly believe such magnificent formations existed in Australia; they seemed far too wild and spectacular to be real, at least to a boy from the largely cleared and subjugated landscapes of the south-west. Since then, like millions of others, I have taken the train trip from Central Station in Sydney up the mountains to Katoomba many times. It is wonderful that the people and successive governments have resisted the temptation to mine, log, clear and otherwise destroy this outstanding natural treasure.

I was an advisor for the Jenolan Caves Reserve Trust for many years and that gave me outstanding opportunities to explore and enjoy the variety of features of the area. The caves are in excellent condition for their age and have taught us a great deal about karst formations and cave environments. They have also provided many generations with significant enjoyment. I had my first taste of adventure caving later in life than most. I was taken to try out the Plughole Cave, where there was a fairly tight squeeze called the S-Bend, which challenged my slightly claustrophobic tendency. That I managed to get through this tiny tube, with arms in front and mobility provided by scrabbling toes, still seems amazing. Real speleologists would see this as child's play, no doubt, but it was a great achievement for me. And after all, overcoming fears is an important part of many wilderness experiences. Another memorable component of my visits was the pleasure of encountering a superb lyrebird on one of the many walking tracks.

PATH TO WORLD HERITAGE

THE BLUE MOUNTAINS are arguably the home of the Australian conservation movement. The national parks movement began here and there was a strong focus on the idea of wilderness from Myles Dunphy and others. The establishment of the Blue Mountains National Park in 1959 was the first of many steps towards the present million hectares of protected World Heritage. Along the way, there were challenges from those who would rather have mined the coal seams interspersed with the sandstone layers and others who proposed a range of transformations. But protection won out and eventually ensured that the integrity of these natural ecosystems was maintained.

The initial World Heritage nomination by the Australian Government gave emphasis to the evolution and diversification of the unique eucalypts. In a comprehensive assessment, the IUCN suggested that, if the principal focus was eucalypts, then a serial nomination (like the Australian Fossil Mammal Sites) would embrace the widest possible field. Such a site might include parts of the Australian Alps and some of the south-west of Western Australia. It was also noted that there were already elements of the eucalypt story in existing World Heritage areas. On this advice, the World Heritage Committee deferred the listing in 1999 and invited the Australian Government to consider a serial listing approach, among other matters.

The next meeting of the World Heritage Committee decided to inscribe the Greater Blue Mountains Area World Heritage site, against the continuing advice of IUCN to defer the nomination, based on the persuasive advocacy of the Australian Government. The Australian delegate at the time indicated that there would be further work on a series of sites for a full cover of the eucalypts and that this would be undertaken under the new federal legislation (*Environment Protection and Biodiversity Conservation Act 1999*). To date, this has not been done. A number of concerns have been raised about the protection of the site, notably in relation to human activity, and in particular to mining activities adjacent to the area, the proposal to raise the Warragamba Dam wall and the encroachment of inappropriate recreation and tourism infrastructure and activities. The New South Wales National Parks and Wildlife Service, which manages the site, closely monitors these concerns, investigating—and mitigating—their potential to cause harm. After the construction of a new airport in western Sydney was announced in 2014, the Australian Government conducted an extensive environmental impact survey, which was then provided to the committee, reassuring UNESCO that the development was unlikely to adversely affect the site's outstanding universal value. Other regular reports, as required by the convention for every World Heritage site, keep the committee informed about the status of immediate and ongoing threats. One significant event was the bushfires in the summer of 2019–2020 that saw 739 906 hectares impacted, or 71 per cent of the property. Over 200 plant and animal species have been affected by the bushfires, with more than 140 million mammals, birds and reptiles impacted across the property.

Some plant communities and species, such as the many species of eucalypt, have special fire-adaptive traits, and the heath shrub has been observed to be sprouting new populations from seeds in burned landscapes. Other species and habitats, such as the Wollemi pine, are highly restricted and more sensitive to the impacts of fire. This means that parts of the Greater Blue Mountains Area and its species will recover at different rates.

A comprehensive suite of monitoring and recovery actions, including control of feral animals and weeds, and waterway rehabilitation, is in place to track the ongoing health and recovery of areas and species and to build their resilience.

Significant rain and flooding in eastern Australia in 2020–2022 have also impacted the property and affected recovery programs. These recent impacts form part of the emerging effects of climate change on our World Heritage properties.

OPPOSITE: Myles Dunphy and Bert Gallop at Jenolan Caves in 1912.

PURNULULU NATIONAL PARK

Millions of years ago, a plateau of ancient sandstone rose perhaps 700 metres above the surrounding terrain. Warmth and wetness were all it took to dissolve some of the silica holding it all together. Then, quartz sand grains were picked up by rainwater gushing across the plateau. Where the layers of rock hold some moisture near their edges, organisms thrived, and in so doing created a dark grey crust that held in moisture and sediment. In areas without that crust, water flushed out sediments, leaving a rusty orange rock face exposed. Gradually, the crevices cut down through the underlying material, steeply and deeply incising the rock to form sheer chasms. At the top of the formation, this process left rounded edges and cones, beautifully banded, their colours changeable in the sunlight. In the deep crevices, water excavated hollows under the rocks, forming pools of water for plants and animals.

LOCATED IN THE KIMBERLEY REGION of far north Western Australia, Purnululu National Park covers 239,723 hectares of outstanding wilderness incorporating the distinctive beehive formations of the Bungle Bungle Range. It is this spectacular and extensive formation of cone karst, with its colourful banding and smooth conical shape, that has attracted international attention and that led to World Heritage inscription. The Bungle Bungle Range came to national attention quite recently compared with other sites, with the first aerial photograph published in 1982 to immediate acclaim. Purnululu displays the full extent of these magnificent landforms, created by a unique combination of geology, erosion and cyanobacteria.

The distinctive cones formed over an extremely long period, taking some 20 million years of erosion for the original quartz sandstone to take its current form. The cyanobacteria that helped create these cones are the same type of bacteria that created the stromatolites in Shark Bay (see page 127). The non-marine cyanobacteria of Purnululu form dark grey horizontal crustal bands only a few millimetres deep but of varying widths up to several metres and it is these that give the Bungle Bungles their remarkable appearance. Some bands have been traced for several kilometres. A much smaller version of this geomorphological feature forms part of the Greater Blue Mountain World Heritage Area in New South Wales.

PREVIOUS PAGES: The spectacular rock formations at Purnululu rise above the flatter surrounding terrain.

OPPOSITE: The horizontal banding on the Bungle Bungle Range is created by cyanobacteria, the same type of bacteria that formed the stromatolites at Shark Bay.

ABOVE: *Livistona* palms at the entrance to Echidna Chasm

I first became aware of Purnululu in 1984 when reading a book called The Land Beyond Time, which featured John Olsen's paintings of the north-western part of Australia. As a postscript, a chapter was added on the Bungle Bungles and the text and images were immediately appealing, describing an expedition in the wet season. Like many other Australians, I felt there was something magical about this fantastic landscape with its unique structures and I was very keen to visit. I finally made the journey in 2001, before it became a World Heritage site. My wife and I drove our four-wheel drive on the track from the Great Northern Highway to the campground and had glimpses of the rocky outcrops on the way in. Once we arrived, we were astonished by the conical beehive formations. We hiked some of the tracks, continuously amazed by the core features: banded karst cones, gorges and amphitheatres. As we walked along the various drainage lines and narrow chasms, the scale of the Bungle Bungles was hard to comprehend. A helicopter flight gave a very different perspective of the spectacular views and convinced my wife she should have been a chopper pilot. I was interested to discover later that some park managers believe such aerial visitation might be less problematic for the environment than ever-increasing ground visits.

The impressive Ord River runs through the southern part of the area and provides the south-eastern boundary of the national park, flanked on the north-west by Osmond Creek, a tributary of the Ord. Access to the park is challenging, with one relatively rough four-wheel drive track leading to the small area of campsites. While the focus of the World Heritage listing is the Bungle Bungle Range, with its unique visual splendour, there are another three distinct habitats that make up the environment and are a part of its outstanding universal value. The Ord River valley is a grassy, riverine ecosystem with open forests along the eastern and southern borders of Purnululu. Broad sand plains separate the uplands from the black infertile soils and open woodland of the river country. To the west are more extensively wooded limestone ridges. The combination provides habitat for an overlap of desert and savanna life forms—17 different vegetation communities include open woodland and spinifex, with many eucalypts, grevilleas and wattles. So many species of spinifex (of the genus *Triodia*) grow in the park that it is considered a centre of diversity and endemism for this unique Australian grass. In the gorges, there are closed forest communities—extensions of northern monsoonal forests—containing palms, ferns and orchids. Over 600 plant species have been recorded in the national park.

Animal life also reflects this mix of tropical and desert species, including 298 vertebrate species, of which 149 are birds. Some of the birds at the southern end of their normal range, such as the pheasant coucal, channel-billed cuckoo and spangled drongo, are associated with monsoon vegetation and more likely to be seen in the palm communities of the gorges and riparian forests further north. Others show their dry-country affinity, such as the zebra finch and budgerigar, preferring the desert environment. Specialised locals prefer the rocky outcrops, including great bowerbirds and the sandstone shrike-thrush, whose melodic calls resonate in the chasms and gorges. The desert mouse

AT THE SOUTHERN END OF THE PURNULULU NATIONAL PARK VISITOR AREA, A SHORT WALK GIVES ACCESS TO THE MAGNIFICENT CATHEDRAL GORGE. HERE, EROSION HAS NOT ONLY CARVED OUT A CHASM BUT ALSO A MASSIVE NATURAL AMPHITHEATRE, COMPLETE WITH WATERHOLE AND, AT TIMES, EVEN WATERFALLS.

and the nocturnal burrowing skink reach their northern limit in Purnululu, while the distinctive large-footed mouse-eared bat is at its most inland occurrence here.

It is predominantly the area's geological features that are considered to be of outstanding universal value. Within the 45,000 hectares of cone karst are many extraordinary elements, few of which are accessible to the public except by foot. An impressive example is Echidna Chasm in the north of the park. Stands of *Livistona* palms provide a green contrast to the orange sandstone, but colours within the canyons reflect the sun's angles and change dramatically throughout the day. The narrow chasm, only a metre wide at one point, opens to the sky above 200-metre-high walls. At the southern end of the Purnululu National Park visitor area, a short walk gives access to the magnificent Cathedral Gorge. Here, erosion has not only carved out a chasm but also a massive natural amphitheatre, complete with waterhole and, at times, even waterfalls. There are many other gorges and waterholes throughout the ranges, undoubtedly important resources for the Gija and Jaru peoples—the Traditional Owners of the area.

PREVIOUS PAGES: Cathedral Gorge.

OPPOSITE TOP: Purnululu, photographed here from Kungkalahayi Lookout, features a diverse range of spinifex species.

OPPOSITE BOTTOM: The sandstone strike-thrush is just one of some 149 species of bird that live at Purnululu.

PATH TO WORLD HERITAGE

IN 1982, FILMMAKER GUY BASKIN took the first aerial images of the Bungle Bungles when filming his television series *The Wonders of Western Australia*. The unique quality of the area was immediately recognised and the Western Australian Government set up the Bungle Bungle Working Group with conservation, mining, tourism and Indigenous representatives. The report from this group triggered a study by the Commonwealth House of Representatives Standing Committee on Environment and Conservation, which recommended in 1985 that 'the Australian Heritage Commission in consultation with the Australian National Parks and Wildlife Service and relevant Western Australian State Authorities undertake investigations to establish the national and international significance of the Bungle Bungle region'. In 1987, the state government declared the area a national park, following the establishment of ties with the area's Traditional Owners and extensive work to reduce the impact of feral animals, especially cattle and donkeys. Subsequently, the state and Commonwealth governments agreed to make a formal nomination of the area for World Heritage listing.

As soon as the area became more widely known, there was an immediate interest in it as a tourist destination. A local family from nearby Timber Creek started offering trips to see the area in 1985 and numbers gradually increased using the rough access track. By 1996, there were 14,500 people driving in and another 40,000 taking scenic flights. Improvements to the road saw an increase to about 30,000 driving visitors per year in 2022. Since inscription, much work has focused on establishing cooperative management arrangements with Traditional Owners, but final outcomes await settlement of Native Title and other matters. There has been serious disruption and dislocation of the Indigenous community but despite this they retain a continuing association with the area. In the World Heritage nomination documents there was strong reference to Indigenous cultural heritage, including such examples as paintings, stencils, engravings, artefacts and shelters:

> *The rock art of the East Kimberley is yet to receive the research attention given to other World Heritage properties, such as Kakadu National Park. However, in 1988 a three month survey of archaeological sites on the northern and western margins of the Bungle Bungle Range recorded over two hundred separate sites within the boundaries of Purnululu National Park. The paintings depict a range of animals including crocodiles, turtles, fish, kangaroos and emus as well as human stick and snake-like figures ... The sites often include stencils, mostly in red ochre, of hands and in at least one instance feet, including those of children as well as implements such as boomerangs and spear throwers.*

The site was nominated under cultural heritage criteria and endorsed as such by the UNESCO advisory body ICOMOS, but complexities with legal and management structures for Indigenous involvement led to deferment of the cultural heritage inscription. Members of the Jaru and Gija peoples make up a majority on the Australian World Heritage Advisory Committee and some areas of the park are designated as Indigenous living areas. Recognition and protection of Indigenous cultural heritage is a significant focus of management and both federal and state governments support the idea of co-management but await a formal initiative from Traditional Owners. It is likely that the formal question of a cultural heritage listing will be taken up at a future date.

A low-altitude aerial view of the Bungle Bungle Range shows extraordinary detail of the range as a whole.

ROYAL EXHIBITION BUILDING AND CARLTON GARDENS

Standing in the middle of Carlton Gardens today, the view to the south is of high-rise buildings in Melbourne's city centre. What might we have seen 250 years ago? On Kulin Country, the view is of land falling away from a slight rise, through a woodland of eucalypts and wattles, to the bank of a significant river, the Birrarung (Yarra River), perhaps 100 metres wide. Kangaroos and wallabies graze in grassy areas, and the air is alive with bird calls. Depending on the season, there are families taking fish from the river or people participating in larger ceremonial gatherings. One hundred years after the arrival of the British, another ceremonial gathering took place on this spot. Flags waved in downtown Melbourne to mark 1 October 1880 as a public holiday. Thousands assembled to watch a procession of more than 5,000 soldiers and sailors, tradesmen and fire brigades, march along Collins and Spring streets, row after row, block after block. To the sound of a 17-gun salute, seven carriages ferried the vice-regal party to the front of the magnificent new exhibition building, a marvel of architecture and construction, exhibiting art and industry from around the world.

JUST A FEW HUNDRED METRES from Parliament House in downtown Melbourne, Carlton Gardens provides the setting for the Royal Exhibition Building. Occupying 26 hectares of prime real estate, this World Heritage site was designed and built to accommodate the great Melbourne International Exhibition in 1880. The scale, design and grandeur of the building reflect the values attached to industrialisation. It was conceived as a palace of industry, a place to both show and celebrate the wondrous achievements of science and the industrial revolution. All manner of novel machinery, products, inventions and art from across the world were the focus of such international exhibitions and had enormous appeal for visitors. The building was meant to be visually splendid and even today it remains impressive.

There were many international exhibitions in the second half of the nineteenth century, beginning with the Great Exhibition of London, held in the purpose-built and magnificent Crystal Palace in 1851. Although there had been exhibitions of products and achievements since the eighteenth century, these were somewhat akin to regional fairs of an earlier era in Europe, and so the 'exhibition era' is usually dated from the London exhibition with its strong sense of empire and colonialism. Each international exhibition set out to chart material and moral progress within the world context. Grand exhibitions were held throughout the world, in New York, Paris, Dublin, Vienna, Amsterdam, Boston, Edinburgh, Barcelona and Brussels, but also in the colonial outposts of Santiago, Cape Town, Kingston, Sydney, Melbourne and Dunedin.

The Melbourne exhibition building was designed by English architect Joseph Reed. Born in Cornwall, Reed travelled to Melbourne in 1853 at the age of 30. He established an architectural practice and was subsequently commissioned to design many city buildings, including the State Library of Victoria, Melbourne Town Hall, Victorian Trades Hall

PREVIOUS PAGES: Detail of the Royal Exhibition Building ceiling beneath the central dome.

LEFT: Looking across the Carlton Gardens lake to the front of the exhibition building

and Wesley Church. His design for the Royal Exhibition Building drew heavily on European traditions, including the Italian Renaissance style with a Florentine dome. The building is cruciform in floorplan and incorporates features of earlier exhibition buildings: great portal entries, viewing platforms and towers. Reed also redesigned the gardens in conjunction with the exhibition building and so they form an integral part of the overall site.

The spacious interior of the building was designed for displays, with the timber flooring of the ground floor a deliberate feature. Originally made from Baltic pine, it was replaced initially with Victorian ash and later cypress pine, and has most recently been restored using spotted gum, to magnificent effect. During an exhibition, it is only partially visible, seen along the thoroughfares between stalls where visitors walk from one end to the other. Entering the building through one of the majestic porticos, visitors immediately notice the breathtaking ceilings, the balconies and balustrades of the second floor and then, beyond, the central dome's cupula. Looking closer, the decorative elements become apparent: mythical characters appear beneath the central dome; underneath the building's arches, sylphs represent the four seasons, Morning, Night, Truth and Justice; on the arches, lunettes representing themes of peace, war, federation and government face the central dome. Some of these Classical Victorian decorations have been revised over time and now reflect different themes to those presented in 1880, which included allegorical images promoting the arts, science, industry and agriculture, along with the coats of arms of the exhibiting nations. Before the current motifs, a scheme that emphasised Empire and glory was painted over the original for the centennial exhibition in 1888. The current designs were painted in 1901.

ABOVE: The original exterior of the exhibition building, showing extensive rear structures that no longer exist.

OPPOSITE: A plan of the gardens for the 1880 international exhibition also shows the extent of the original exhibition building. The section marked 'Main Building' comprised 100,000 square metres of temporary annexes built specifically for the 1880 exhibition and then subsequently removed.

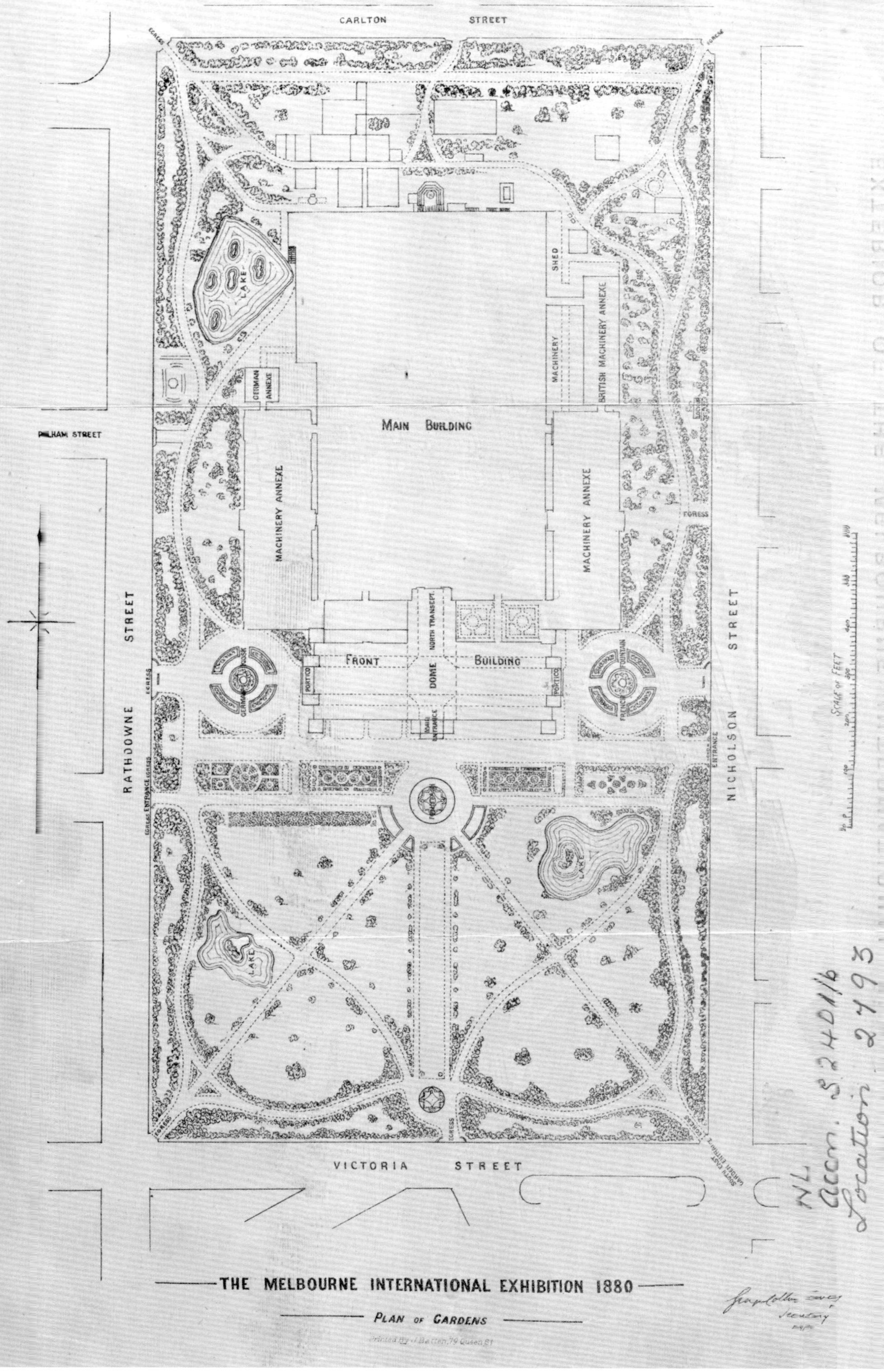

THE MELBOURNE INTERNATIONAL EXHIBITION 1880

PLAN OF GARDENS

Printed By J Batten, 79 Queen St

The world responded enthusiastically to the Melbourne exhibition, with 33 counties and colonies represented in a range of major and minor courts, including the states of Australia. A great diversity of products was on display, from the finest jewellery to huge working machines of industry and warfare. More than 12,000 different exhibitors brought over 32,000 exhibits, showcasing everything from art to industry, from domestic goods to the latest agricultural and industrial processes and achievements. Exhibition areas were divided up by country, with space at a premium. In addition to the various courts of the main building, temporary sheds were constructed to accommodate everyone. Separate galleries of art were a feature, as were many operating machines showing the latest technology, such as cotton-weaving equipment and a French machine producing bonbons on demand. Precision equipment, including microscopes and musical instruments, was on display, as was a full-scale Krupp field gun. Each of the dominant European countries displayed extensive sets of wares and the total floor space for exhibits came to about 84,000 square metres. More than 1.3 million visitors were recorded during the 212 days of the official exhibition (1 October 1880 to 30 April 1881).

The southern portico faces the pleasure gardens designed for the International Exhibition of 1880. The vista across the gardens encompasses Melbourne city, today a cluster of

Victorian locomotives on display for the 1880 exhibition.

high-rise buildings but originally with a much lower profile. Carlton Gardens was set aside in the formative stages of the Melbourne settlement by Charles Joseph La Trobe, then superintendent of the Port Phillip District. In 1839, he envisaged a green belt surrounding the city, of which the Carlton Gardens would be a part. It functioned as a town common for livestock grazing and firewood until 1855, when its management was transferred to the City of Melbourne. Its final design, in association with the International Exhibition, was accomplished by William Sangster and Joseph Reed. In the foreground of the gardens is the magnificent ten-metre-high Exhibition Fountain, designed by sculptor Joseph Hochgurtel. Hochgurtel's fountain, depicting four youths dancing below symbols of the arts, science, commerce and industry, reflects the themes of the exhibition. The gardens have a further two fountains, as well as two lakes and a host of pathways connecting the many trees and flowerbeds. The south garden remains much the same as it was originally conceived and provides a magnificent setting for the building. With sweeping lawns and both European and Australian plantings, the gardens are of aesthetic, architectural, social, historical and scientific significance. Less survives of the north garden because the layout has changed multiple times since the 1880s, as structures have been added and removed. Some avenues of mature trees do remain.

As the only remaining purpose-built exhibition building that survives in its original form and function, the Royal Exhibition Building has international significance. Other great exhibition buildings have been lost completely, such as London's Crystal Palace (destroyed by fire in 1936), or remain as a fragment only, such as the buildings in Chicago and St Louis. The Eiffel Tower, now part of the Paris, Banks of the Seine World Heritage Area, was not an exhibition building but rather was built as an exhibit in the 1889 Universal Exposition of Paris.

MORE THAN 12,000 DIFFERENT EXHIBITORS BROUGHT OVER 32,000 EXHIBITS, SHOWCASING EVERYTHING FROM ART TO INDUSTRY, FROM DOMESTIC GOODS TO THE LATEST AGRICULTURAL AND INDUSTRIAL PROCESSES AND ACHIEVEMENTS.

ABOVE: The Hochgurtel Fountain.

FOLLOWING PAGES: The building's Florentine dome lends old-world charm to the modern Melbourne skyline, seen peeking over the roofline.

I have had two revelations about international exhibitions thanks to this World Heritage site. The first is the way in which the philosophy, design and function reflect the colonial and industrial perspective of the time. The exhibitions gave pleasure to visitors and promoted trade, informing about the frontiers of practical knowledge that were transforming society and creating a new, more interconnected world. In Melbourne, much of the flavour was about giving Australia a place in the modern world, no longer simply part of a British Empire. But it also coincided with the rise of social concern about the questionable benefits of industrialisation for workers. The Dickensian interpretation of the Victorian-era divide between rich and poor was in full swing and support for a wholly celebratory approach to industry was soon to end. Such sites of cultural heritage help focus our minds on our origins and achievements and perhaps prepare us to think about a better future.

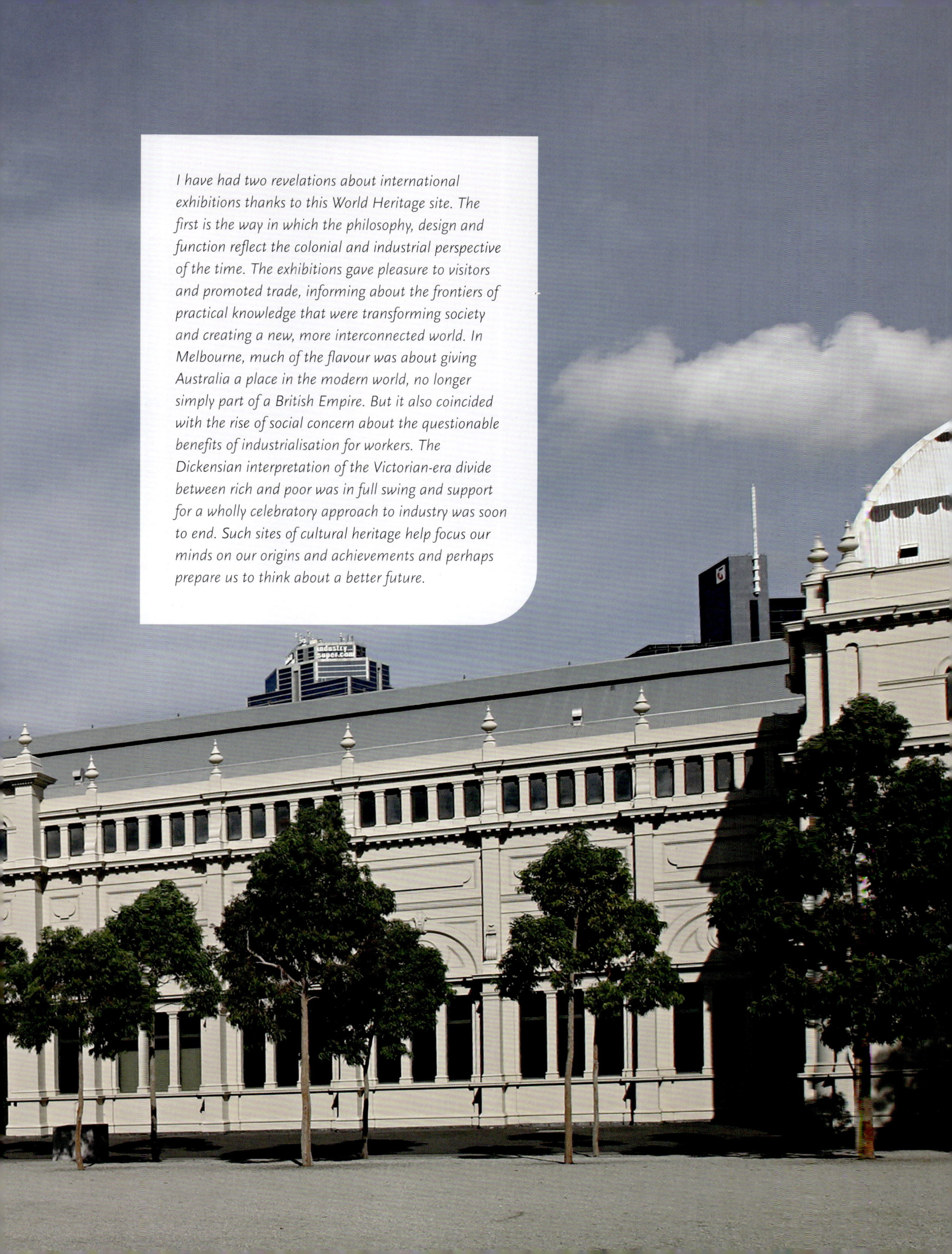

Although in many instances exhibition facilities were not meant to survive permanently, the Royal Exhibition Building was built to last and indeed was used for a second great international exhibition in 1888—the Centennial Exhibition, celebrating 100 years of European settlement in Australia. Most significantly, perhaps, in the Australian context, the building was used for the first sitting of the new Commonwealth Parliament on 9 May 1901. The historic scene was memorialised in a massive painting by Tom Roberts. For the following 25 years, the Victorian Parliament sat in the western annex of the exhibition building while the federal government occupied the Victorian Parliament House as it awaited new accommodation in Canberra. The building also continued to house many kinds of exhibitions, as it does today. It was used as an emergency hospital ward in 1919 to cope with 1,800 patients affected by the deadly Spanish flu. During the Second World War, it became temporary troop accommodation and, afterwards, a migrant reception centre, as well as having many other short-term uses. Both building and grounds continue to be used for many purposes. Since the early 1900s it has been used to house students taking examinations. During the Covid-19 pandemic it was used as a vaccination centre, creating a nice synergy with the first global pandemic.

Some changes to the site have been significant, including the loss of a substantial part of the north gardens as well as alterations within the building over the course of its life. However, the World Heritage Committee considered that the site had retained enough integrity in its overall intent and functioning to be inscribed as World Heritage. Subsequent conservation plans seek to maintain that into the future.

Since the site's inscription on the World Heritage list in 2004, a number of projects have been undertaken to preserve and reinstate key elements of the site from the international exhibitions period. Recent projects have included reinstatement of the parterre garden beds, restoration of the Western Forecourt and reinstatement of the Dome Promenade.

A monotype of Tom Roberts' impressive 'Big Picture' shows the opening of the first Australian Parliament, in the Royal Exhibition Building.

PATH TO WORLD HERITAGE

DESPITE SOME MODIFICATIONS, the function of the Royal Exhibition Building and Carlton Gardens continued long after they were first constructed and they remain in excellent condition. The building's significance was first recognised in 1982, when it was listed on the Victorian Government Buildings Register. A conservation analysis undertaken in 1983—in the wake of a decision to demolish remnants of the 1880 machinery hall—resulted in an ongoing process of reinstatement and restoration. The building was transferred to the Victorian Heritage Register in May 1993 (the Carlton Gardens were added in 2002). Both building and gardens were listed on the Register of the National Estate (maintained by the Australian Heritage Commission) and when the *Environment Protection and Biodiversity Conservation Act 1999* led to the establishment of the National Heritage List, the site was reassessed for inclusion, which offered further protection.

The federal government, in conjunction with the Victorian Government and the Melbourne City Council, submitted a nomination for World Heritage in 2002 but ICOMOS recommended the World Heritage Committee defer inscription. There was some uncertainty about the authenticity and integrity of the site, with ICOMOS expressing reservations about the impact of the new museum building. The Australian Government undertook extensive additional analysis and provided the committee with a supplementary set of information in June 2004. The World Heritage Committee inscribed the site on 1 July 2004.

Prior to its World Heritage inscription, restoration work on the building reinstated missing ornamentation along the parapet.

SYDNEY OPERA HOUSE

Imagine listening to Beethoven's magnificent Fifth Symphony in the Concert Hall of the Sydney Opera House with more than 2,000 other people. This space has been shared by many performers, not just the resident Sydney Symphony Orchestra. Above the stage, the high, vaulted ceiling rises 22 metres through a layer of 18 acoustic reflectors, past the world's largest mechanical action pipe organ to the astonishing giant sails. Here, we sit within one of the greatest indoor performing spaces on earth. Beneath us the foundations contain a raft of other rooms and facilities, enabling the full complexity of modern entertainment and artistic support. Even further below on the Bennelong peninsula, jutting out into the harbour waters, there are fragments of the past from before Europeans, from the days of the Gadigal people of the Eora Nation. This land has seen gatherings of many peoples through time, sharing experiences, joyous voices and the sounds of music. It brings a diversity of culture together, linking ancient lives with modern creations, all within an exceptional architectural monument.

THE SYDNEY OPERA HOUSE was inscribed on the World Heritage List in 2007 as a masterpiece of twentieth-century architecture. Covering just 5.8 hectares, it is the most recognisable building in Australia. In nominating the site, the Australian Government noted that the design by brilliant Danish architect Jørn Utzon 'represents an extraordinary interpretation and response' to the breathtaking setting of Sydney Harbour. Opened in 1973, the building is made of three sections of interlocking, shell-like vaults that house seven performance spaces and a restaurant. Centred on a large terraced podium, the complex continues to function as a venue for diverse performances.

The history of the Sydney Opera House is almost as famous as the design. In 1956, the New South Wales Government, led by Premier Joseph Cahill, announced an international competition to design a new opera house and appointed an independent jury to assess entries. The competition brief was very open and neither set extensive parameters nor provided a cost limit. The design required, essentially, two performance halls, one for opera and the other for symphony concerts. Over 900 architects registered and 233 submitted designs (all anonymous). The outstanding concept drawings submitted by 38-year-old Utzon were chosen unanimously, a decision the Australian Government described in its World Heritage nomination as 'unexpected, bold and visionary'. At that stage, no structural engineer had been consulted and it was not clear whether these design ideas could actually be put into practice. Novel engineering techniques and materials were required and it was fortunate that an excellent partnership developed between Utzon, engineer Ove Arup and others to bring the Opera House to fruition over the following 16 years and beyond. It was not plain sailing. Disagreement over cost, and media criticism, led the new Robert Askin government to force out Utzon when the building was only partly completed. Fortunately, the continuing work by architect Peter Hall mostly followed Utzon's original design parameters, enabling its successful completion. Utzon was re-engaged as Sydney Opera House architect in 1999 and has since completed new works that have brought great benefits to the building.

PREVIOUS PAGES: Sydney Opera House as seen from the water.

OPPOSITE: Shadows criss-cross the harbour-side foyer of the concert hall.

ABOVE: An aerial view of the world-famous sails shows the distinctive pattern of the pearly white roof tiles.

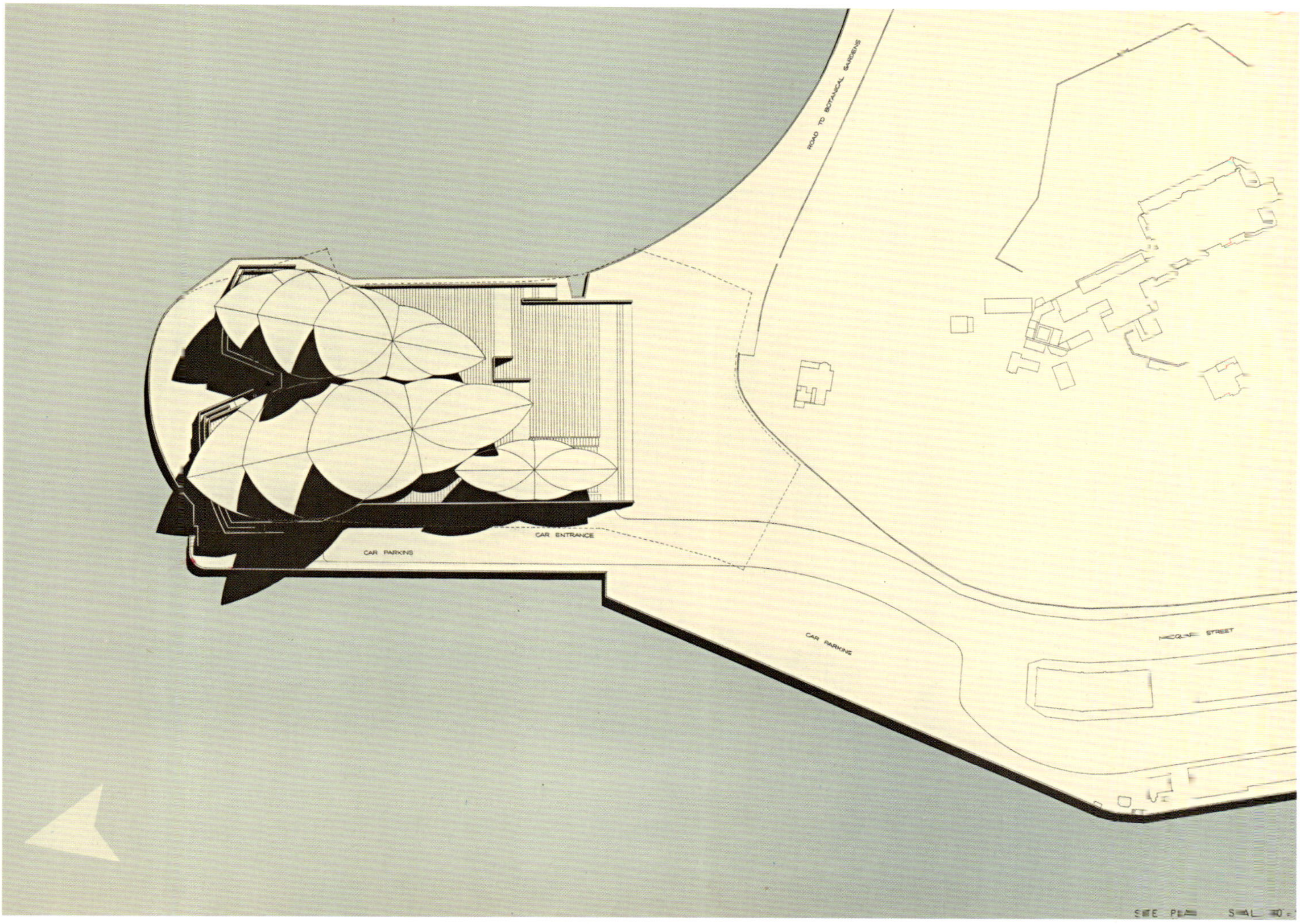

Utzon's original plans for the Opera House design competition, showing east (opposite, top) and north (opposite, bottom) elevations, as well as the plan for the whole site (above).

The nomination document offers a poetic description:

> *[The Sydney Opera House] is a daring and visionary experiment that has had an enduring influence on the emergent architecture of the late 20th century and beyond ... The forecourt is a vast open space from which people ascend the stairs to the podium ... The podium steps, which lead up from the forecourt to the two main performance venues are a great ceremonial stairway nearly 100 metres wide .. The final shape of the shells was derived from the surface of a single imagined sphere ... [They] are faced in glazed off-white tiles while the podium is clad in earth-toned, reconstituted granite panels ... The outstanding success of the building today is in large part due to the way it is grounded in the site: 'It appears to rise out of the ground as a landform almost as though it were a geological extension of the earth itself'.*

The site of the building, Bennelong Point, protrudes into the harbour from the botanic gardens just east of the Sydney Harbour Bridge, guaranteeing a beautiful setting. People have gathered here for thousands of years. Shell middens found in the area confirm that it was used by local Indigenous people before Europeans arrived. The peninsula, known as Tubowgule by the Gadigal, is a place where generations of Aboriginal people have feasted, sung, danced and told stories. The British, who arrived in 1788 to colonise the area, used Tubowgule to confine their cattle and horses, and renamed it as Cattle Point. Colonists crushed middens from around the harbour, including Tubowgule, and mixed the remnants with water to make a lime slurry to use as mortar. The promontory thus became known as Limeburners' Point. The present name honours Woollarrawarre Bennelong, an Aboriginal man who was kidnapped by Governor Phillip and became a trusted advisor on traditional ways, a diplomat and peacemaker.

Utzon's design intentionally lifted the performance halls well above the surrounding environment. The tops of the vaulted shells soar above ground level and the terraces function as a stepped access to the entrances. This approach is partially inspired by Mayan architecture, in which living areas are elevated eight metres or more above the surrounds. The terraces create a podium from which there are views across the harbour. The two main halls sit alongside each other, their long axes running more or less north–south. The larger was meant to be the opera hall but the Australian Broadcasting Commission argued for it to become the symphony concert hall, with the smaller hall being re-designated for opera. There are over 1,000 rooms in the complex, almost all of which are beneath the podium, including all the control facilities for the performance halls. After he was re-engaged, Utzon redesigned the reception hall, celebrating rather than hiding the ceiling's concrete beams and designing a bright, wall-length tapestry to enhance acoustics.

The Sydney Opera House continues to be one of the busiest performing arts centres in the world, each year staging up to 2,500 performances and events for 1.5 million performance patrons and an estimated 10.9 million site visitors. Beneath the sails, seven performance venues host an amazing range of events beyond the opera and symphony, including dance, music, theatre and talks. In a novel development the vaulted shells are now used as screens, the special ceramic tiles lit at night for many kinds of presentations. A spectacular example is Badu Gili, a daily six-minute lightshow of artwork by Indigenous artists, celebrating and honouring Indigenous culture.

OPPOSITE: Looking out into the audience in the Concert Hall.

FOLLOWING PAGES: The Sydney Opera House is a commanding presence along the shoreline of one of the world's most stunning harbour cities.

Like many other Australians, my first view of the Sydney Opera House was from an aeroplane coming in to land at Sydney airport. The beauty of the building and its setting took my breath away—photographs just did not do it justice. It helped that it was one of those gorgeous Sydney days with blue skies and even bluer waters. Since then, I have made many visits to the Opera House and each time I rediscover just what an amazing building it is. In November 2011, when I was a member of the National Landscapes Reference Committee, I was privileged to be taken on a walking tour around The Rocks area and immediate vicinity of the Opera House by a young Indigenous guide.

We talked about the former middens on the peninsula and inspected the original rock foundations of the Harbour Bridge and other areas where some of the material contained shells from the old middens. It seemed to me to create a positive connection between the original Australian people and this fabulous modern cultural icon, recognised globally as distinctively Australian. Some people see a visual link between the design of the Opera House and the shape of sea shells, another connection with the middens. I was struck by the symbolism of this heritage place also: the world's most outstanding modern building linked to the world's oldest living culture.

PATH TO WORLD HERITAGE

NO SOONER HAD the Sydney Opera House been officially opened by Queen Elizabeth II on 20 October 1973 than plans were formulated for appropriate international recognition. In 1980, Australia submitted a World Heritage nomination dossier referred to as 'The Sydney Opera House in its Setting', including the Sydney Harbour Bridge and the surrounding waterways of Sydney Harbour from Bradleys Head to McMahons Point. ICOMOS recommended deferring the listing, partly because of the difficulty in assessing a building that had only so recently been completed. They suggested that inscription should be 'deferred until [the Opera House's] exemplary character or its role as model [appeared] more clearly attributable to the creation of Jørn Utzon'. Reflecting on this, the World Heritage Committee 'considered that modern structures should only be accepted when there was clear evidence that they had established, or were outstanding examples of, a distinctive architectural style'. There was interest in seeing a revised nomination for Sydney Harbour, based on the first European settlement in Australia, that might include the Opera House and the Sydney Harbour Bridge, but not as primary elements.

In the following 25 years, much was written about Utzon and the outstanding quality of the building. The building continued to develop, and various alterations and additions, following Utzon's design principles, enabled the authenticity and integrity of the building to be retained. In assessing the 2006 nomination, ICOMOS was fully supportive of inscription. Its recommendation to the committee concluded that 'the Sydney Opera House stands by itself as one of the indisputable masterpieces of human creativity, not only in the 20th century but [in the] history of humankind'.

In 2018, the New South Wales premier instructed the Sydney Opera House Board to advertise a horse race on the Opera House sails, to emphatic community protest.

This is in stark contrast to the Hiroshima Prefecture's approach to Hiroshima Peace Memorial World Heritage Area. As part of a beautification plan, nearby building owners avoid advertising on the walls of buildings facing the World Heritage site. Although possibly extreme, this indicates what might be expected of countries in respecting heritage. There is an illumination of the sails policy as part of the Sydney Opera House Conservation Management Plan for the site.

OPPOSITE: Queen Elizabeth II at the official opening of the Opera House on 20 October 1973.

ABOVE: A public festival on the date of the official opening included fireworks and F-111 swing-wing bomber displays over the harbour.

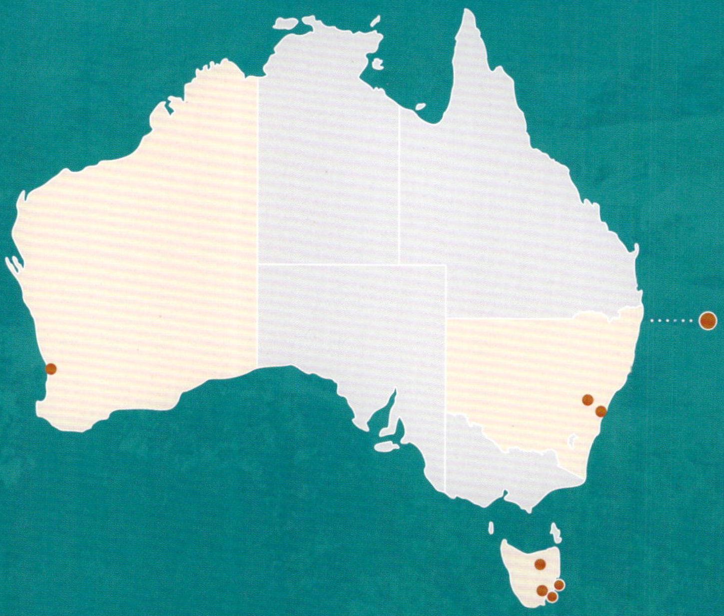

AUSTRALIAN CONVICT SITES

Sentenced to transportation for seven years! A desperate theft of food translated to forced migration to an entirely new world. Convicts waited in jail before joining others bound for Botany Bay. What followed was eight months of diabolical conditions sailing across the oceans: locked between decks, sealed tight in rough weather amid the bilge and vomit of the other 100 or more wretches. Eventually, the ship put ashore in Sydney, if lucky, or Van Diemen's Land if not. Some were forced into chain gangs to perform hard labour, others given to freemen, masters who could be kind or cruel. Women, treated like cattle, were given to men or put to work in factories or as servants. Men were the source of most labour in the colony. All were locked away each night. Living in harsh conditions, every complaint was dealt with by flogging—100 lashes for insolence. Those who re-offended might be shipped off to the hell on earth that was Norfolk Island, where few survived and there was no hope. Many died at every stage of transportation, but most survived the brutality and eventually became the founders of a new society.

THIS SERIAL WORLD HERITAGE PROPERTY consists of 11 convict sites dating from eighteenth- and nineteenth-century colonial Australia. They housed male and female convicts, including children, transported from Britain and Ireland. Everything about this system was harsh: the penalties imposed by the British justice system; the appalling conditions of the prisons (often hulks) before transportation; the arduous and long journey to Australia; and the harsh treatment at all stages. On arrival, convicts faced punitive imprisonment and forced labour, sometimes for the government, sometimes for private citizens. Over 3,000 sites across the country were constructed to hold the new arrivals: in total some 166,000 convicts passed through these sites between 1788 and 1869. The Australian Convict Sites are exceptional examples of Australia's convict history, a unique record of global convictism and associated beliefs and practices about crime and punishment.

Initially, the British system of harsh punishment was meant to deter criminal behaviour but it was subsequently linked with a philosophy of reformation of criminals that also provided cheap labour to developing colonies. These, at times contradictory, objectives were the focus of much debate throughout the transportation and convict era. Transportation to Australia was a penalty for both hardened criminals and those convicted of relatively minor offences, including women and children from the age of nine. Some were political prisoners, some simply wanted to form a union of workers. The areas making up this serial listing illustrate different types of convict sites, a range of living conditions experienced by convicts and various types of forced labour they endured. Convicts contributed substantially to the European settlement of Australia, both in the forced labour required of them and in their subsequent activities as free settlers. Each of the 11 sites contributed to one or more of three thematic elements around the forced migration of convicts: expanding the British geopolitical sphere of influence; punishment and deterrence; and reformation of convicts.

PREVIOUS PAGES: The view from Scorpion Hill, Port Arthur, taking in the church ruins.

OPPOSITE: The Cascades Female Factory in 1900.

The 11 sites make up a diverse architectural ensemble of more than 200 structures, ruins and remains. There are buildings for housing, confining and managing convicts (including penal stations); female factories; a juvenile prison; underground and solitary cells; barracks; stockades; hospitals; and churches. The sites also encompass convict-built infrastructure, such as roads, dockyards, a colliery, crank mills, kilns, brickworks, agricultural properties, government houses and administrative buildings. Between them, they capture a wide range of punitive horrors, uncivilised living conditions and attempted reformation. While the scale of transportation to Australia was far greater than in any other penal colony in the world, many other European countries administered similar systems and so the sites represent an international practice of forced migration of convicts that lasted almost a century.

Each of the 11 convict sites offers a significant window into Australia's convict past and helps us appreciate the physical reality as well as the initial and evolving philosophy behind its existence. Only some of the sites retain the appearance of the past (some are ruins) making interpretation a key element of understanding

ABOVE: The remains of the main penitentiary building at Port Arthur, with the guard tower visible to the left and the hospital up the hill on the right.

OPPOSITE: The coachman's cottage at Brickendon Estate in 2007.

MACHINE SHOP
5

Kingston and Arthurs Vale (Norfolk Island)

The more than 40 buildings at this site were built from locally quarried limestone and local timber, with the convict quarters near the coast and the supporting military and administrative facilities overlooking them from the ridge behind. This site is in excellent condition and provides evidence of the essential elements of convict transportation. The worst offenders and reoffenders were sent here and its punitive reputation was extreme—the treatment was intended to instil fear in lawbreakers in Britain and the colonies to prevent future offences. The crank mill for grinding maize is a good example of the priority given to punishment over redemption. Convicts spent long hours turning the crank mill's heavy wheel. It was a physically and mentally punishing task that left little energy for learning to lead a productive life in their new environment. The incidence of suicides, riots and executions was far higher than at other penal stations and Norfolk Island gained an international reputation as 'hell on earth' and more recently as a prototype for concentration camps.

Old Government House and Domain (Parramatta, New South Wales)

This extensive landscape includes five buildings and many archaeological remains, and covers 37 hectares of parkland. It was the headquarters for the administration of the colony under 12 successive governors. Many convicts and soldiers were also stationed there. Female convicts worked as servants and male convicts undertook agricultural and industrial tasks. Important decisions about the colony and its inhabitants were made in Government House, which has today been restored. It is significant chiefly because it represents a reformative rather than purely punitive approach to the treatment of convicts.

Hyde Park Barracks (Sydney, New South Wales)

Designed for 600 people, the main building initially served as accommodation for convicts who worked offsite during the day. The three-storey barracks was built in the Georgian style, with a timber-shingled roof. Its large, central clock reminded convicts of their time-bound routines of work and religious services. Their lives were confined to the main and supporting buildings by a three-metre-high stone wall. A major new interpretation program has been developed for on-site visitors.

Old Great North Road (Wisemans Ferry, New South Wales)

This site is a 7.5-kilometre-long section of the 250-kilometre Great North Road, constructed between 1828 and 1835 by male convicts. The hard labour it took to build the road was part of the punishment for reoffending convicts and was designed to reinforce the fear of transportation while also creating transport infrastructure for the colony. Thousands of sandstone blocks were quarried by hand from cliffs and shaped into massive retaining walls, spillways, gutters, culverts and intricate drainage systems, most of which remain today. This extraordinary feat of colonial engineering was made possible by convict labour. Various other structures also endure, such as temporary housing for convicts.

Cockatoo Island (Sydney Harbour, New South Wales)

The largest island in Sydney Harbour at 18 hectares, Cockatoo Island is a sandstone plateau upon which substantial convict-built infrastructure supported quarries, grain storage and prisons for secondary offenders, as well as a large dry dock. It was a place of harsh labour. The 13 convict-built structures are constructed from local sandstone blocks and include prisoner barracks, a hospital, mess hall and kitchen, and various military quarters.

The entrance to the machine shop on Cockatoo Island in 19[illegible]7.

Brickendon and Woolmers Estates (Longford, Tasmania)

The Brickendon farming estate (420 hectares) includes 20 buildings that continue to function as they did when part of the convict working environment. This is a well-preserved example of a site where convicts were assigned to a private individual (in this case, the Archer family, who still own and run the property today) for the purpose of useful work and reformation. Wheat and barley have been grown continuously on the property since convict days and the landscape remains much as it was, including 30 kilometres of convict-planted hawthorn hedge fences. The neighbouring Woolmers Estate (formerly owned by the Archer family but now owned by a private trust) includes 18 buildings and structures on 13 hectares. Many of the function-specific buildings are still in excellent condition on both estates, including chapels, blacksmith shops, stables and various convict quarters.

Darlington Probation Station (Maria Island, Tasmania)

The buildings and ruins at Darlington reflect the philosophy of close surveillance of convicts to ensure rehabilitation while also applying psychological pressure to conform. This reflects the 'probation system' which replaced the initial 'assignment system', designed to ensure both punishment and rehabilitation of the convict. The civil and military administrators lived close by, with functioning buildings reminding the convicts that hard labour, regulation, religious instruction and schooling were the keys to reform. Punitive components included solitary cells (for short-term punishment), separate apartments for the worst behaved and barracks for the remaining convicts.

Cascades Female Factory (Hobart, Tasmania)

Three of the original five Female Factory compounds remain, mostly as ruins, showing the features of this segregated site that sought to accommodate, punish and reform female convicts. The site was almost self-

The main cell block at Fremantle Prison in 2007.

sufficient, with a hospital, nursery, laundries, cookhouses, administrators' apartments, offices, solitary cells, various workshops, stores and a church. Various classes of convicts were treated differently, according to a probation system in which convicts who reformed received progressively better treatment. The site, at the bottom of a cold valley, was well hidden from Hobart's other citizens.

Port Arthur
(Tasmania)

Comprising more than 30 convict-built structures and ruins on 136 hectares of relatively undisturbed and attractive landscape, Port Arthur initially focused on punishment and later on reform. A four-storey penitentiary sits at the centre of the other, lower buildings. The site operated as a major industrial complex, supporting agricultural work, and dockyard and shipping activities, as well as timber, quarrying and dairy activities. A boys' prison on adjacent Point Puer is a reminder that some convicts were children and highlights the attempt to manage/punish different classes of convicts differently. Over 1,000 people are buried in the convict-era cemetery.

Coal Mines
(Little Norfolk Bay, Tasmania)

The relatively undisturbed bushland site of 214 hectares contains the remains of more than 25 substantial buildings as well as the remnants of a coal mine. Convicts provided the labour to mine the coal and operate the preparation and loading facilities at jetties. The site also includes many of the buildings associated with accommodation and control of the convicts. Underground and other solitary punishment cells, separate apartment cells, prisoner barracks, officers' quarters, workshops, other facilities and the chapel are evident, as are houses and other buildings. The working conditions for the convicts were appalling: inadequate air and clouds of coal dust created a stifling underground environment, where they were forced to spend long hours with little respite.

Fremantle Prison
(Western Australia)

The convict barracks and prison at Fremantle were built by convicts and constructed of locally hewn limestone blocks; the perimeter wall is an impressive six metres high. Within it are eight buildings, including cell blocks, chapels, punishment cells, a hospital and workshops. Outside the walls are guard facilities and houses for warders and others, including soldiers and priests. The prison was created towards the end of the convict era and is associated with the abolition of transportation. It became a penitentiary for Australian prisoners.

* * *

Each of these sites has distinctive elements that illustrate the different treatments of convicts in Australia. They vary both in the nature of their purpose and in the type of remains within the varying geographic and management regimes. Collectively, they provide an exceptional testament to the global history of convict transportation and treatment. Covering a period of 80 years, the sites also demonstrate developments in, and changes to, theories about the transportation and treatment of convicts.

My family has frequently enjoyed playing and singing folk songs, and among our favourites is 'Botany Bay'. When I was young, I gave little thought to its background as we sang the rollicking ballad around a campfire and it was not until much later that I came to appreciate the details of transportation. One insight came from seeing some of the heritage sites that allow us to better imagine the conditions under which the convicts were held. The extremely small spaces for solitary confinement, without light or comfort, must have been highly oppressive and indeed many convicts became mentally affected by these conditions. It was meant to be a harsh deterrent and the records of suicide suggest the reality was brutal. Some of the sites were designed with elements for deliberate punishment or torture, including grain-grinding treadmills. A second insight came from the writing of Marcus Clarke, whose novel For the Term of His Natural Life *clearly captures some of the horrific conditions that convicts endured. The idea of transportation and punitive internment as a deterrence sickens me, whether as part of our history or in the present time. It has no place in a civilised society.*

PATH TO WORLD HERITAGE

GIVEN THE SCOPE AND COMPLEXITY of the field and the geographic diversity of remains and structures, it is not surprising that the process of nominating this site involved substantial expert studies by both Australian and international parties over many years, as well as a high level of coordination with the Australian Government. These expert studies began in 1995 and continued through to the listing of the sites on the Australian National Heritage List in 2007. The sites were included on the tentative World Heritage List in June 2000. A draft nomination for World Heritage listing was subject to expert peer review and was submitted to UNESCO in 2008.

ICOMOS undertook technical missions in 2009 to evaluate the claims in the nomination document and asked Australia a series of follow-up questions. These sought a more comprehensive statement to justify the serial nature of the nomination (11 sites collectively having Outstanding Universal Value) and details about the process of analysis that decided which of Australia's 3,000 or so convict sites should be included. Satisfied by the Australian Government's responses, ICOMOS advised the World Heritage Committee that the serial property has Outstanding Universal Value, including integrity and authenticity considerations. On the basis of advice from ICOMOS, the committee made a number of recommendations for Australia to consider in its ongoing management of the sites, particularly concerning respect for landscape values and integrity; the removal of anachronistic structures at some sites; effective participation from local site committees in management of the site as a whole; and the publication of monitoring indicators.

OPPOSITE: The remains of a room housing the treadmill at Norfolk Island, 1953.

ABOVE: Gallows Gate at Norfolk Island, 1953.

NINGALOO COAST

Imagine a large whale shark swimming south from Indonesian waters as part of her annual migration. In the open ocean she may dive deeply, sometimes down 1,000 metres, seeking concentrations of larval fish and crustaceans. Mostly, she will stay near the surface, following a cryptic trail to find food. In the course of just a fortnight, she might travel 1,000 kilometres through the Indian Ocean from the shallow coastal waters of Java to Ningaloo, where she can feed on coral spawn and inadvertently entertain hundreds of tourists. A researcher may one day affix a monitoring tag to this gentle giant to learn about her life when she leaves this World Heritage area, her home for just a few months each year.

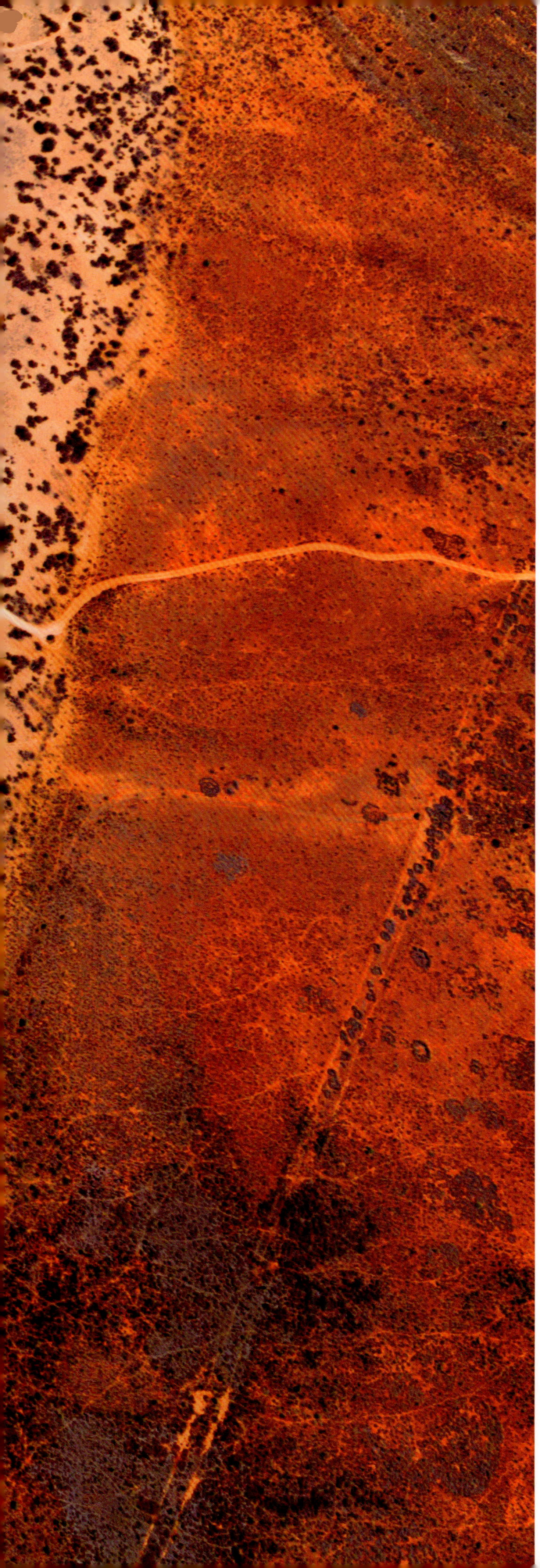

ON THE OPPOSITE SIDE of the continent to the Great Barrier Reef, there is another exceptional reef system, inscribed as a World Heritage area in 2011 in recognition of its outstanding natural phenomena and beauty, as well as its astonishing biodiversity. Stretching 290 kilometres along the westernmost coast of Western Australia, the Ningaloo Coast site contains one of the world's longest fringing reefs. The complex nearshore reef follows the coastline for 200 kilometres, never more than seven kilometres offshore and at times as close as 200 metres. In some places, such as Coral Bay, the corals grow within a few metres of the sandy shore. But this World Heritage area also incorporates the terrestrial environment of Cape Range with its karst formations, cave life and other associated habitats and species along the arid coastline. Altogether, the site has a high number of land-based species native to this area and high marine species diversity and abundance.

For much of its length, the coast is backed by sand dunes averaging eight metres high, aligned by prevailing winds in a north–south orientation at the Northwest Cape but more north–west to south–east along most of the coast. In places, the east–west drainage lines of streams and rivers have cut through the Cape Range, creating canyons and gorges that flow down to the coast and dissect the offshore reef. Where they occur, sandy beaches provide ideal conditions for turtle nesting and as many as 10,000 nests appear each year, mainly belonging to green turtles but also loggerhead and hawksbill turtles. Each species is present here in globally significant numbers and they appear to enjoy remarkable breeding success in this remote location, thanks to good management, especially the control of foxes over recent years.

PREVIOUS PAGES: Ningaloo Reef.

LEFT: An aerial view of the Cape Range National Park and Ningaloo Marine Park shows the vivid, and starkly contrasting, colours of the desert and ocean.

EACH YEAR, MIGRATING HUMPBACK WHALES PASS THROUGH NINGALOO ON THEIR WAY NORTH TO CALVING GROUNDS OFF THE KIMBERLEY COAST AND THEN AGAIN ON THEIR RETURN TO ANTARCTIC WATERS.

One of the most remarkable marine attributes of Ningaloo is the annual aggregation of the world's largest fish, the magnificent and beautiful whale shark. Each year between March and August, following the annual mass coral spawning and seasonal nutrient upwelling, some 300 to 500 whale sharks gather at this unique and accessible environment. The Ningaloo Reef is remarkably rich in wildlife, with over 300 species of coral, over 700 species of reef fish and more than 600 species each of mollusc and crustacean. Marine plants are also well represented with more than 1,000 species identified. Apart from the six species of marine turtle recorded in the area, there are many ray and shark species, as well as several open ocean fish species that live close to the coast, such as mackerel, marlin and sailfish. Each year, migrating humpback whales pass through Ningaloo on their way north to calving grounds off the Kimberley coast and then again on their return to Antarctic waters. Many other cetacean species have been recorded here, including Bryde's, minke and southern right whales, as well as dolphins. Although less well studied, the life on the ocean floor uncovered by recent research includes a high number of sponge species and 25 species of echinoderm new to science.

Over 700 species of fish make their home among the corals of the reef.

Yardie Creek has carved a gorge through the rocky landscape of Cape Range National Park.

In contrast with the blues and greens of the Ningaloo marine environment, the terrestrial Cape Range National Park is arid brown, grey and red. Much of the biological treasure is hidden beneath the surface in the hundreds of caves, conduits, pools and aquifers that form part of this karst limestone environment. The surface is rugged—the dry valleys and collapsed caves have sharp, rocky edges. Fluted rock outcrops protect gorges, springs and depressions of various kinds from erosion. Numerous species of arthropod and two endemic fish species live in the subterranean aquatic environments. These highly specialised cave animals embody long-term evolutionary responses to the changing geology and biogeography of the region. The Cape Range karst system features a number of landlocked water bodies that have underground connections to the ocean (anchialine systems). The majority of anchialine animal species here are not found elsewhere in the southern hemisphere and are unrelated to communities in other karst environments in Australia. The IUCN evaluation of the area concluded that the existence of relict rainforest fauna and highly specialised endemic aquatic invertebrates within the same cave system is exceptional: this area has 'the highest cave fauna (troglomorphic) diversity in Australia and one of the highest in the world'.

The above-ground terrestrial environment forms an area of desert, recognised for its high level of species richness and endemism, particularly of birds and reptiles. Shrubby trees like eucalypts and wattles dominate, beneath which grow numerous shrubs, spinifex and seasonal wildflowers. This ecosystem supports around 200 bird species, including emus, parrots, honeyeaters and finches, as well as some isolated subspecies such as the grey shrike-thrush and spinifex pigeon. An isolated subspecies of the western bowerbird is found only at North West Cape. Over 90 reptile and amphibian species live on the peninsula, of which some are at the limits of their habitat range and others are confined to the area. These all add to the natural history significance of the World Heritage area, and are of interest but are not of outstanding value compared with the marine and subterranean animal and plant life.

ABOVE: The seasonal wildflower Yardie morning glory is named after Yardie Creek.

FOLLOWING PAGES: Whale shark at Ningaloo Reef, 2018.

I first visited Ningaloo in 1996 with my colleague Alastair Birtles. Our purpose was to work with local tour operators to develop suitable protocols for a swim-with-whale-sharks program that would better ensure sustainability. For me, it was an astonishing introduction to these gentle giants. Over the course of several days we tried to document the likely impacts that swimmers would have on the animals. On one occasion, a young male whale shark followed me around for nearly 30 minutes, swimming slowly and close behind me as I backstroked in circles until too tired to continue. The whale shark stopped vertically near the vessel, apparently curious. Usually, a whale shark would simply swim onward with its mouth open, ingesting food as it swam. At the end of the trip, we proposed a number of measures, including limiting how close both swimmers and vessels can be to whale sharks, which were adopted by tourist operators and later by the park management agency. They have since become more sophisticated, and ongoing research seeks to guarantee protection of the species and ensure the encounters are sustainable. Today, the whale shark management is considered world's best practice.

PATH TO WORLD HERITAGE

NINGALOO WAS ADDED TO Australia's National Heritage List on 6 January 2010. The nomination for World Heritage listing was prepared by the Australian Government in the same year, with cooperation from the Western Australian Government. This is a rare case of a relatively smooth path to nomination, largely due to the cooperation between the state and Commonwealth governments and the lack of substantial local controversy. It was also the result of community consultation over several years in the lead-up to National Heritage listing and to World Heritage nomination. This helped allay concerns expressed by local graziers and others about how the listing might affect their operations. The main point of contention was the boundary and resultant large size of the property. The extension of the area from the existing marine park to cover terrestrial areas provoked some concern and drove extensive community and commercial consultation with Indigenous people, graziers, mining interests, commercial tourism operators and fishing interests. A peer review and reference group also assisted the government in preparing the nomination. This was unusual at the time but may become the norm as the committee gives increasingly more weight to technical and scientific evidence in assessing site nominations.

The IUCN provided the World Heritage Committee with a report that was itself based on the findings of 11 external reviewers and extensive consultation during its field mission to review the veracity of the nomination. The nomination stated that the site met three criteria for inscription and the IUCN supported the claim for two—for superlative natural phenomena and significant conservation sites—but also noted that the site could meet the criteria for outstanding ecological and biological processes. When inscribing the property, the World Heritage Committee also accepted the IUCN's recommendation to delete several terrestrial areas included in the nomination. These were referred back to

Yardie Creek.

Australia for further consideration in collaboration with the owners of these lands, with the proviso that they could be added to the property later as a minor boundary change. It also recommended other actions, including that Australia consider renominating the property under the ecological and biological processes criterion; consider including Exmouth Gulf within an expanded site; bolster its innovative volunteer programs; and strengthen the collaboration with the Western Australian Department of Fisheries.

Western Australia's Department of Conservation, Biodiversity and Attractions manages the area, and the World Heritage Committee suggested that Australia use existing and successful management models—such as that exemplified by the Great Barrier Reef Marine Park Authority—to enhance the management framework and capacity of the Ningaloo Coast World Heritage Area. While this suggestion has not been implemented, an agreement between the state and national governments has seen the establishment of the Ningaloo Coast World Heritage Advisory Committee, which brings together research, conservation, tourism, Indigenous and local communities to advise on management.

A recent IUCN conservation outlook report suggests that climate change is the 'greatest threat' to this site, noting that 'The coral reefs within the Ningaloo Coast appear to have escaped the coral bleaching event which affected many other coral reef areas in 2016; however concerns are still high regarding their future.' The Ningaloo Coast site remains well protected. It has an effective range of management plans and qualified staff, but concerns remain about potential future threats from already observed and predicted climate change impacts, future oil and gas or coastal development activities.

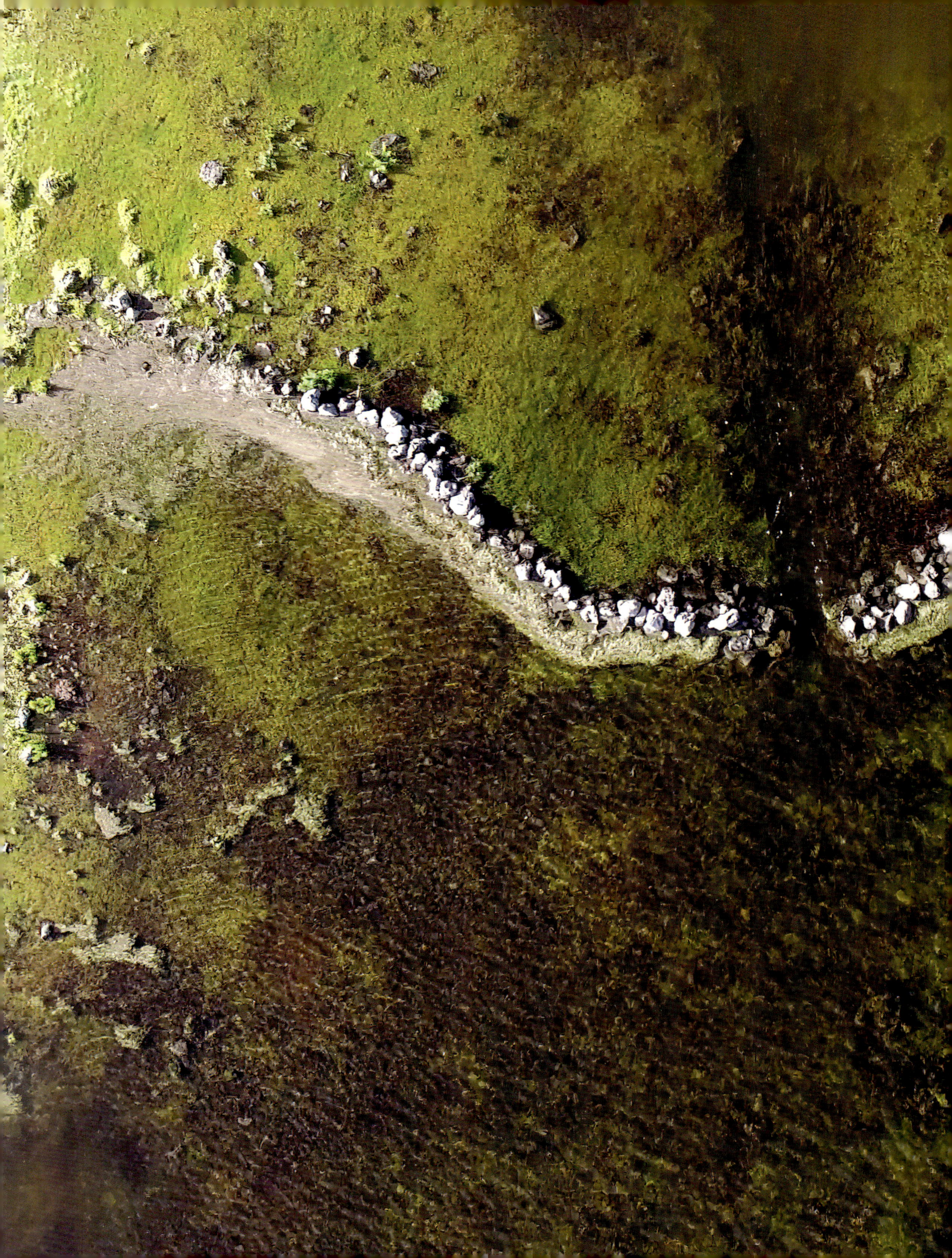

BUDJ BIM CULTURAL LANDSCAPE

Some 37,000 years ago on Gunditjmara Country, straddling the western Victoria coastal plains, an extremely important event transformed the landscape. A volcanic eruption produced the domed hill of Budj Bim and the very extensive lava flows that blocked the prior drainage lines, forming an arc of basalt all the way to the coast and beyond. In doing so, Budj Bim created new waterways, extensive wetlands and associated rock fields, which the Gunditjmara people used to build an ingenious living aquaculture system, the world's oldest. Budj Bim also enabled relatively permanent living spaces associated with the eel-farming: the people constructed stone houses and villages. The story of Budj Bim is part of the oral history, the creation stories, of the Gunditjmara who have lived here for aeons. This story is supported scientifically by recent accurate estimates of the age of the basalt and is a fascinating example of landscape evolution, with the oral history record coinciding with the geological story. The ancient nature of the eel-farming (at least 6,600 years) is simply exceptional and that is why Budj Bim became the first place to be listed on the Australian National Heritage List and subsequently to be inscribed on the World Heritage List in 2019.

THE BUDJ BIM CULTURAL LANDSCAPE tells a story of people, place and nature that is part of the living and ancient history of this area of south-eastern Australia, the traditional Country of the Gunditjmara people. It is now known that people were living in this part of western Victoria at least 37,000 years ago, based on the discovery of a stone axe below dated tephra material at Tower Hill, just 47 kilometres southeast of Budj Bim. What is quite fascinating is that the story of the Budj Bim volcanic eruption is presented in the oral history of the Gunditjmara people, suggesting an astonishing link between ancient volcanic events and the persistent occupation of this country by the Traditional Owners. The most recent volcanic activity of this small western Victorian section of the Newer Volcanic Province seems to have occurred within a short period; all the dated vents seem to have erupted within the same period around 37,000 years ago. In the deep-time story of the Gunditjmara, Budj Bim was the creative being who changed the country and thereby enabled the development of aquaculture and settled living conditions for the long-term benefit of the people.

In the Dreaming, the ancestral creators gave the Gunditjmara the resources to live a settled lifestyle. They diverted the waterways, and gave us the stones and rocks to help us build the aquaculture systems. They gave us the wetlands where the reeds grow so that we could make the eel baskets, and they gave us the food-enriched landscape for us to survive.

Eileen Alberts – Gunditjmara Elder

PREVIOUS PAGES: One of the weirs in the Budj Bim Cultural Landscape.

OPPOSITE: An aquaculture channel and pond.

ABOVE: Budj Bim World Heritage Ranger Tyson Lovett-Murray holds up a *kooyang* (short-finned eel).

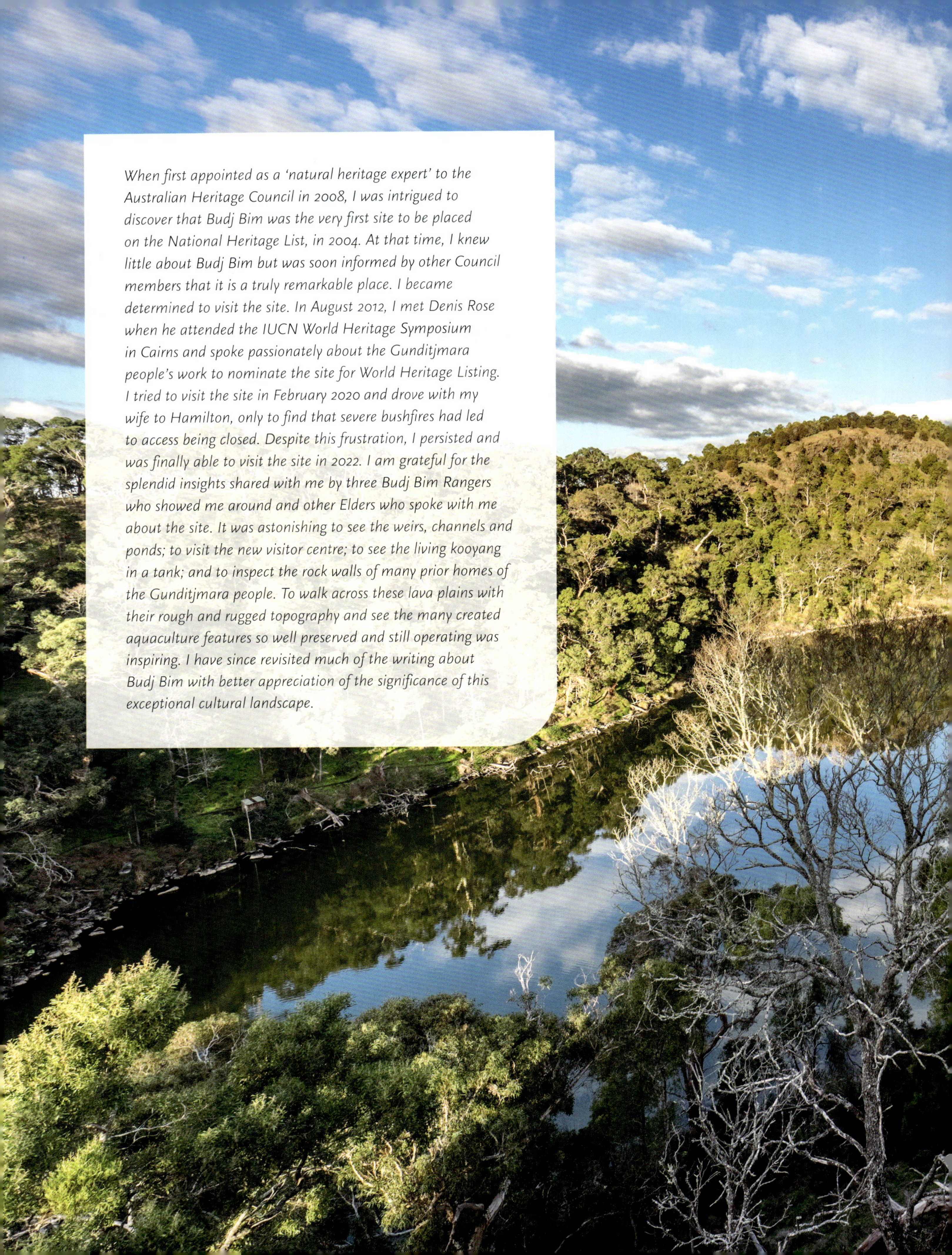

When first appointed as a 'natural heritage expert' to the Australian Heritage Council in 2008, I was intrigued to discover that Budj Bim was the very first site to be placed on the National Heritage List, in 2004. At that time, I knew little about Budj Bim but was soon informed by other Council members that it is a truly remarkable place. I became determined to visit the site. In August 2012, I met Denis Rose when he attended the IUCN World Heritage Symposium in Cairns and spoke passionately about the Gunditjmara people's work to nominate the site for World Heritage Listing. I tried to visit the site in February 2020 and drove with my wife to Hamilton, only to find that severe bushfires had led to access being closed. Despite this frustration, I persisted and was finally able to visit the site in 2022. I am grateful for the splendid insights shared with me by three Budj Bim Rangers who showed me around and other Elders who spoke with me about the site. It was astonishing to see the weirs, channels and ponds; to visit the new visitor centre; to see the living kooyang in a tank; and to inspect the rock walls of many prior homes of the Gunditjmara people. To walk across these lava plains with their rough and rugged topography and see the many created aquaculture features so well preserved and still operating was inspiring. I have since revisited much of the writing about Budj Bim with better appreciation of the significance of this exceptional cultural landscape.

The lava from Budj Bim flowed down the prior drainage systems, filling and overflowing the valleys. The resulting wetlands included large lakes (Tae Rak) and swamps (Condah Swamp) and many ponds, as well as flowing sections of rivers and streams braiding across the disrupted landscape. It is within these new waterways that the *kooyang* (short-finned eel, *Anguilla australis*) subsequently established a habitat and, along with rich wetland plant resources and other wildlife, became the key natural resources for the Gunditjmara.

Kooyang have a fascinating lifestyle, and while they do occur along the southeast coastal rivers and wetlands of Australia, they breed only in the distant Coral Sea. Mature males and females of various ages and sizes (up to a metre or so in length) leave their freshwater homes and swim perhaps 3,000 kilometres from Budj Bim, travelling about 30 kilometres per day. They swim deeply during the day (700–900 metres deep) and more shallowly at night (100–300 metres). Once they have arrived at the spawning site —the location of which is not precisely known—they mate, and females might lay between 500,000 and 3 million eggs before dying. After hatching, the young leaf-shaped eels drift with sea currents until eventually they arrive offshore along the east and south coast of Australia, where they enter the freshwater streams and estuaries as glass eels and elvers. They then swim upstream in their thousands and after arriving at the perfect habitat of the Budj Bim Cultural Landscape spend the next 15–20 years growing and fattening ready for their final adult journey back to the spawning site.

It is this huge population of growing eels that provided an opportunity for the Gunditjmara people to establish a sustainable fishery, sufficient to support a sedentary lifestyle. But first the physical construction of the waterways and ponds, trapping areas and water-flow controls had to be built using the basalt stones that formed part of the landscape. The same stones were used to build houses for more permanent human occupation close to the fishery. The aquaculture system designed and created by the Gunditjmara goes back a staggering 6,600 years (at least). The fish management involved women weaving special elongated eel traps (*gnarraban*) that were up to three or four metres long and would be placed in gaps designed for that purpose in the weirs constructed across the built channels. The eel traps were woven from a local *Carex* species of grass (*puunyaart*), and were long enough to prevent the eel from turning around. Alongside this very effective method, some eels were also speared by hunters.

THE BUDJ BIM CULTURAL LANDSCAPE WAS INSCRIBED ON THE WORLD HERITAGE LIST UNDER TWO CRITERIA: AS AN EXCEPTIONAL TESTIMONY TO THE CULTURAL TRADITIONS, KNOWLEDGE, PRACTICES AND INGENUITY OF THE GUNDITJMARA PEOPLE; AND AS AN OUTSTANDING REPRESENTATIVE EXAMPLE OF HUMAN INTERACTION WITH THE ENVIRONMENT.

Some of the eels were smoked after being caught; partly burned-out trunk hollows of large trees were used for this purpose. Smoked eels were well preserved and could be used for trade.

The Budj Bim Cultural Landscape was inscribed on the World Heritage List under two criteria: as an exceptional testimony to the cultural traditions, knowledge, practices and ingenuity of the Gunditjmara people; and as an outstanding representative example of human interaction

PREVIOUS PAGES: Lake Surprise.

OPPOSITE: Remnants of a stone house along Darlots Creek near Tae Rak.

with the environment. The serial site incorporates intact and outstanding examples of the largest aquaculture complexes and a representative selection of the most significant and best preserved smaller structures. The three aquaculture elements are Tae Rak in the north, and Kurtonitj and Tyrendarra in the south. At each site there are channels, weirs, dams and ponds that contribute to the overall complexity and operation of the Gunditjmara aquaculture system. The waterflows, once reduced by draining in the 1950s for farmland, have now been restored and the system is in good condition and is visible in the landscape.

Two key features of the northern component are Budj Bim itself, the volcanic source of the lava flow that created the landscape, and Tae Rak, the extensive lake that formed as a result of the lava flow. From the Budj Bim National Park campground today, the lookout enables a view across the water-filled crater of Budj Bim (Lake Surprise) in the foreground to the volcanic vent with its domed head and stone teeth (*tung att*)—a vista remarkably consistent with a painting from the same point completed by Eugene von Guérard in 1865. It was from this point that the Tyrendarra lava flow emerged and spread in a huge arc west and south, flowing all the way to the coast of that time, which today means 15 kilometres of the flow is beneath the present sea level. The second key feature is Tae Rak (sometimes called Lake Condah), a large water body formed when the flowing basalt lava blocked the prior creek that then backed up northward to what is now Condah Swamp. Eventually, water flowed around the barrier and cut a new channel south, called Killara by the Gunditjmara and Darlot Creek by the Europeans. The complex flows of water across, through and beneath the landscape presented opportunities for the Gunditjmara ancestors to create a sustainable wildlife resource involving not only eels, but also fish species, birds and plants. But it was the ingenuity involved in developing the eel aquaculture that is today seen as exceptional, contributing as it did to the development of settled villages of stone-walled houses close to eel harvest ponds and weirs.

The central component of the Budj Bim Cultural Landscape is called Kurtonitj, part of the Budj Bim Indigenous Protected Area, and is an extensive area along the left bank of Killara, some five kilometres south from Tae Rak.

A channel running alongside Tae Rak.

The landscape here is all on the Tyrendarra lava flow and includes water flow channels, weirs, ponds and sink holes. Once a focus for ongoing aquaculture, today it houses interpretive features that enable visitors to better appreciate the aquaculture system and associated activities.

A further five or six kilometres south is the final element of the Budj Bim Cultural Landscape World Heritage Area: Tyrendarra. Stretching between the Killara creek and further to the Palawarra (Fitzroy River) in the west, this area, like Tae Rak, supported a significant village settlement linked to extensive aquaculture operations. Despite significant records of these Gunditjmara houses and villages, including written descriptions, sketches and maps, within a brief time of European settlement most disappeared, along with the people, many of whom were killed or dispersed. More recently, much archaeological evidence has been uncovered, including the walls of many houses.

The frontier wars in this part of Victoria were savage and repeated the tragic experiences found across the country. Despite Gunditjmara men being recognised as excellent warriors, many died and their families often suffered a similar fate or were removed from their land into prisons or missions. Massacres associated with whalers (in 1833–183[illegible]) and subsequently with European settlers escalated to the point where, in 1842, the conflict was called the Eumeralla War. Within the Budj Bim Cultural Landscape two significant massacres are recorded, including one at Killara, near Tae Rak, where 20 people died from arsenic-poisoned flour. It is estimated that the fierce resistance of the Gunditjmara led to between 300 and 350 Gunditjmara casualties during the war. Throughout the conflict the rugged nature of the landscape enabled the Gunditjmara to continue living on parts of their Country despite the opposition of European pastoralists.

For 50 years after 1867, the focus for survival was the Lake Condah Mission and the establishment of an official Aboriginal Reserve. Despite restrictive rules at the mission, people were able to continue their cultural links to their land, including the aquaculture system. It took enormous effort by the Gunditjmara to continue the fight for recognition and it was not until 2007 that Native Title consent determination was reached by the Australian Federal Court.

A channel with a reconstructed stake.

An eel trap woven from water reeds.

PATH TO WORLD HERITAGE

THE ESSENTIAL STARTING POINT FOR World Heritage listing of Budj Bim is the unbroken lineage of the Gunditjmara people on this country and the ingenious ways in which they developed and maintained a sustainable aquaculture system. As the oral history of this site found strong support in geological and archaeological research, more people began to appreciate the exceptional reality of this cultural landscape. But the formal pathway was not easy, requiring both federal and state governments to support the nomination and facing strong resistance from the federal government at the time.

The idea of nominating Budj Bim for World Heritage listing was entirely a product of the Gunditjmara people, driven by the desire to have the stunning culture of their ancestors recognised by a wider community, not just in Australia. Equally, the work was undertaken by Gunditjmara people and projects. Gunditjmara Elder Denis Rose led the community forums, consensus building and the nomination team over a seven-year period following the rebuilding of the Tae Rak weir and its revitalised operation. A World Heritage workshop was held on Country in 2013 and that reinforced and further invigorated the work of developing support and evidence to make a nomination. While the community came together behind the nomination, and the Victorian government also, there was resistance at the national level.

The breakthrough came with a new Prime Minister, Malcom Turnbull, who joined with Victorian Premier Daniel Andrews to announce formal support during a visit to Budj Bim in January 2017. The Gunditjmara then had the task of rapidly putting together a formal nomination document that in the end is a magnificent testament to all involved. Over 250 images and 230 pages of detailed material was submitted to the World Heritage Centre in January 2018. This was to be seen as the first Indigenous-led World Heritage nomination and the first for Australia that was purely based on Indigenous culture. On 6 July 2019, the dreams of the

Gunditjmara were realised as the World Heritage Committee inscribed this outstanding property on to the list of World Heritage. There were many Gunditjmara in the audience in Baku (Azerbaijan) when the result was announced, and even more at home in western Victoria watching the live feed of the meeting.

As a postscript, it is fascinating to record that the bushfire in 2019 that closed access temporarily also exposed previously hidden cultural artefacts, including a major extension to the Tae Rak southwest fishtrap and an additional cluster of stone houses. The combination of on-ground exploration, multiple science projects and modern technology has the prospect of revealing even more understanding of this special part of the world. The opening of the visitor centre at Tae Rak was a significant step in making this knowledge even more accessible to the wider community.

Swamp wallabies are common at Budj Bim.

The jetty at the Tae Rak visitor experience area.

POTENTIAL NEW WORLD HERITAGE SITES IN AUSTRALIA

HOW CAN WE TELL WHAT PLACES in Australia may be suitable for future World Heritage inscription by UNESCO? The World Heritage Committee long ago established a process that encourages each country to submit what they called a Tentative List: a list of properties that are considered to be of outstanding universal value and therefore suitable for inscription on the World Heritage List. A site must appear on a Tentative List for at least one year before it can be nominated for listing. The World Heritage Committee can then plan its work because it can anticipate future nominations from any given region and make appropriate comparisons across the world. By June 2023, a large number of countries had submitted tentative lists (186 countries out of 194). In total, there are more than twice as many sites on the tentative lists as are already included on the World Heritage List. Larger countries have identified many sites: in 2023 China had 62 sites on its Tentative List and India had 52. Even some relatively small countries had many sites on their lists, including Italy with 31, Egypt with 34, Colombia with 11 and France with 36. Unfortunately, Australia has never cooperated fully with this part of the World Heritage Convention and lags well behind most countries in putting forward a comprehensive Tentative List. At present, only four sites are on the Australian Tentative List: Murujuga Cultural Landscape in Western Australia; Flinders Ranges in South Australia; Great Sandy World Heritage Area, a mainland area of similar habitat to that of the K'gari (Fraser Island) World Heritage Area; and an extension to the existing Gondwana Rainforests of Australia World Heritage Area.

This very limited official list means that one must turn to other sources to discover what our future World Heritage might include. Murujuga Cultural Landscape was nominated for World Heritage by Australia in 2023 and the decision is pending. The other three sites on the Australian World Heritage Tentative List are likely to be the next nominations to the committee. Over the course of many years, other areas have been identified as potential World Heritage sites by various organisations (including, in some cases, government departments) or publications. These are summarised in a recent publication by the Australian Committee for IUCN, *Keeping the Outstanding Exceptional: The Future of World Heritage in Australia*. Most of the sites identified are described below. The Australian Government now has a policy of Free Prior and Informed Consent as part of any future World Heritage nomination.

AUSTRALIAN WORLD HERITAGE TENTATIVE LIST

Murujuga Cultural Landscape

The astonishing Murujuga Cultural Landscape was included on Australia's World Heritage Tentative List in January 2020. With more than one million images in an area of more than 37,000 hectares of the Dampier peninsula in the Pilbara region of Western Australia, Murujuga is home to one of the most significant and diverse collections of petroglyphs in the world. It has the densest known concentration of hunter-gatherer petroglyphs anywhere in the world. The nomination of the site for World Heritage was based on a document prepared by the Traditional Custodians of Murujuga, collectively referred to as Ngurra-ra Ngarli. It was submitted in 2023 by the Australian Government. The decision by the World Heritage Committee is pending but I have no doubt it will be added to the list.

Great Sandy World Heritage Area

Australia's original nomination of the K'gari World Heritage Area included a substantial area of the mainland coast, almost as far south as Noosa. After considering IUCN advice, the committee inscribed K'gari without any mainland component. Subsequently, both the Queensland and Australian governments have agreed to nominate the mainland areas under the tentative name Great Sandy World Heritage Area. It includes the Cooloola section of the Great Sandy National Park, the Breaksea Spit to the north

PREVIOUS PAGES: Sand dunes on Lake Eyre's northern shore, 2009.

of K'gari, Platypus Bay, the Wide Bay Military Reserve and the Ramsar Convention-protected wetland of the Great Sandy Strait (including Tin Can Bay). Sharing many of the attributes of K'gari, it also has the world's oldest coastal dune formation, with eight overlapping layers of dunes representing a history of 700,000 years. Together with rich biodiversity—including a wallum vegetation community and abundant bird diversity (more than 350 species, including a number of migratory waders)—the area has great beauty and is part of an ongoing ecological process. It was added to the Australian Tentative List on 4 January 2010.

Gondwana Rainforests of Australia World Heritage Area (extension)

It has long been recognised that there is capacity to add to the Gondwana Rainforests serial World Heritage site. An IUCN technical evaluation in 1993 called on Australia to consider additions that would both enhance the property and improve its management. The extension is likely to include a number of sites in both New South Wales and Queensland, including the Bunya Mountains National Park in the north (Queensland) and Barrington Tops National Park in the south (New South Wales). The sites identified by the IUCN technical report will likely form the basis of this extension. The proposed extension was added to the Australian Tentative List on 7 May 2010.

Flinders Ranges

The well-known Flinders Ranges geological successions represent an extraordinary window into the major stage in Earth's history described as the 'dawn of animal life'. The sites in this serial property present a geological record of Earth's environments over a period of 350 million years that indicates conditions on Earth that gave rise to animal life. The site includes traces of life from microbial stromatolites to early primitive multicellular life, to the advent of animals as exemplified by the remarkable Ediacara Biota. The evolution of skeletons, limbs, eyes, predation and vertical burrowing during the Cambrian radiation was the origin of many modern animal life forms on Earth today. The Flinders Ranges, part of the traditional lands of the Adnyamathanha, is a beautiful semi-arid mountainous landscape located 500 kilometres north of Adelaide, the capital of South Australia. It was added to the Tentative List in 2021.

* * *

On the following pages are listed a number of other Australian sites that would be appropriate for World Heritage listing.

The Australian Heritage Council during an inspection at Murujuga for World Heritage Assessment at the request of the Federal Government, back in June 2010.

POTENTIAL NEW WORLD HERITAGE SITES

Site	State/Territory	Type	Qualities
Arnhem Land	Northern Territory	Mixed	An area of high cultural and natural value equal to bordering Kakadu, Arnhem Land features largely undisturbed natural landscapes and ecosystems. Critically, it offers substantial evidence of the ongoing occupation and management of the area by the Yolngu people for at least 40,000 years.
Australian Alps	Australian Capital Territory, New South Wales, Victoria	Mixed	This serial area incorporates 11 national parks and nature reserves already inscribed on Australia's National Heritage List. The relatively low mountain ranges combine with Australia's unique flora to produce plant communities different from any other in the world. Where most alpine locations feature conifers, the Australian Alps are known for their eucalypt diversity.
Australian Antarctic Territory	Antarctica	Natural	This utterly unique environment has no existing World Heritage listing, despite being named by the IUCN in 1982 as one of the world's greatest natural areas.
Australian desert and arid zones	Northern Territory, Queensland, South Australia, Western Australia	Mixed	These distinctive arid and desert areas contribute to 20 per cent of the Australian landscape. They include the remarkable Channel Country and Lake Eyre, both previously identified by the IUCN as outstanding national areas. There is an option for a serial listing of multiple sites as an example of Australia's desert diversity.
Cape York Peninsula	Queensland	Mixed	This area is a significant example of tropical savanna: it is less degraded by human activity than anywhere else in the world. Along with significant Indigenous cultural heritage, it has an outstanding range of aquatic ecosystems and exceptional ecological integrity. Its complex landscape matrix demonstrates the coast's past connection to New Guinea.
Coral Sea	Queensland	Natural	This sea is an expanse of deep ocean containing corals, cays, islets, atoll systems, seamounts and pinnacles supporting cold-water ecosystems. It is an important site for cetaceans (at least 27 species of dolphin and whale live here), nesting seabirds and marine turtles.
Houtman Abrolhos	Western Australia	Mixed	A chain of 122 islands and associated reefs, one of which is the most southerly coral reef in the world, it is home to the largest colony of wedge-tailed shearwater seabirds in the eastern Indian Ocean as well as significant breeding populations of other seabirds. The area also encompasses the infamous *Batavia* shipwreck and associated survivor camps, inscribed on Australia's National Heritage List.
Kimberley Region	Western Australia	Mixed	An area of spectacular landscapes, the Kimberley has extensive biological diversity due to isolation from human disturbance. Not only do rare human footprints highlight the extent of human presence in the area, preserved dinosaur footprints attest to Mesozoic-era life.

POTENTIAL NEW WORLD HERITAGE SITES

Site	State/Territory	Type	Qualities
Nullarbor Plain	South Australia, Western Australia	Mixed	One of the world's most extensive limestone landscapes, archaeological sites in this area demonstrate Indigenous settlement for at least 40,000 years. Southern right whales visit the coastline here for their annual calving.
Rowley Shoals	Western Australia	Natural	These three remote coral atolls demonstrate undisturbed coral geomorphology rising steeply from the sea floor. Their shallow lagoons are home to an abundance of marine life, including over 230 species of coral and 680 species of fish. The islands host a number of bird species, including a rare breeding colony of red-tailed tropicbirds.
South-western Australia	Western Australia	Mixed	Australia's only global biodiversity hotspot, this region contains almost 3,000 species of endemic plant and a number of threatened endemic animals.

POTENTIAL EXTENSIONS TO EXISTING WORLD HERITAGE SITES

Site	State/Territory	Type	Qualities
Australian Fossil Mammal Sites	Northern Territory, Queensland, South Australia	Natural	An extension to this site could include other locations that would complete the fossil evidence of the evolution and diversity of Australian mammals, including Murgon and Bluff Downs in Queensland and Alcoota in the Northern Territory.
Great Barrier Reef	Queensland	Mixed	This extension could include parts of the Torres Strait, encompassing not only natural phenomena but also the outstanding cultural heritage of the Torres Strait Islander peoples.
Greater Blue Mountains	New South Wales	Mixed	This area could be expanded to include other areas that demonstrate the rich biological heritage of eucalypt trees, as recommended by the IUCN at the time of listing. It could also include the Sydney basin for its high coastal biodiversity.
Tasmanian Wilderness	Tasmania	Mixed	The Tarkine, remarkable for its Gondwanan cool temperate rainforest (the largest in Australia), would be an obvious addition to this area.

THE FUTURE OF WORLD HERITAGE IN AUSTRALIA

HOW WOULD THE FUTURE BE DIFFERENT if all the sites that have been identified as potential World Heritage areas were to be inscribed on the World Heritage List? If Australia's number of World Heritage sites grew from 20 to 40? In geographical terms, so much more of this unique continent would be recognised as globally significant heritage, meaning better legal protection and better conservation management strategies. The list of sites suggested in the previous chapter would cover all Australia's states and territories and embrace many more remote communities, perhaps bringing new opportunities for local people to craft an economic future around their heritage as they share it with other people. Across the world, World Heritage listing has been associated with a significant growth in tourism beneficial to local economies. Locals, too, are often proud to have their backyard places so respected. Across the world, listing brings people together in united appreciation for our natural and cultural marvels. Having imagined it, how can we achieve it?

Australia has often played a leading role in support of the World Heritage Convention and has also legislated to enable the Australian Government to operate effectively according to the terms of the convention. In early years, it took some testing of the Australian Government's constitutional powers to affirm its essential role and to put limits on a state's capacity to destroy heritage. The *Environment Protection and Biodiversity Conservation Act 1999* was forward-looking in ensuring that the Australian Government could protect national and international heritage. The development of the Great Barrier Reef Outlook Report is another example of effective innovation in the face of growing concern. Such documents also enable Australia to better respond to the periodic reports that UNESCO expects each country to provide about their World Heritage

PREVIOUS PAGES: Traditional weaving at Laura Dance Festival, Cape York, 2005.

LEFT: The north-east hills of the Kimberley, 2013.

areas and to respond appropriately to reports from the committee about the condition of its World Heritage sites. The Australian World Heritage Advisory Committee (involving all existing site management bodies) and parallel Australian World Heritage Indigenous Network, formed in 2002, would be excellent mechanisms to better develop the Australian World Heritage management capacity, if funded appropriately. The committee has produced an online guide to managing World Heritage. However, there is a great deal of diversity in management and resourcing between sites and this aspect of our World Heritage responsibility requires more investment.

The future of World Heritage is not simply a matter for UNESCO or the Australian Government. It also depends significantly on communities being more involved in the processes to 'identify, protect, conserve, present and transmit to future generations' the cultural and natural heritage of Australia. One shortcoming of Australia's current system is the opaque process used to identify future World Heritage in Australia and this may well be at the heart of the failure to develop an appropriate Tentative List. No Australian government has yet implemented the ideal of a comprehensive tentative list, even though this is explicitly called for by the World Heritage Committee and has been reinforced by the national committees of ICOMOS and the IUCN in Australia. In some countries, there are excellent processes for engaging with communities to build tentative lists. India, for example, has a national advisory committee that runs workshops across the country to strategically develop a list based on strong scientific evidence and a rigorous approach to gathering information about potential sites. In Australia, the Australian Heritage Council has been

An adult green python at Iron Range National Park on the Cape York Peninsula.

developing a series of thematic studies of heritage that guide a well-founded comparative analysis to identify its most significant heritage. But even the simple discussion of potential World Heritage is fraught with difficulty in Australia, perhaps partly because World Heritage debates here have more often been seeded with conflict than celebration.

While comprehensive consultation about potential sites is limited, there has been a better effort to engage local people in the management of existing sites. Local engagement and support of a site can be essential to sustainability. Most Australian sites are managed by bodies that include a community consultative committee, a scientific advisory committee and Indigenous representation. This ensures the sites remain connected to their local communities, working with rather than against them. In some cases, the board of management is derived from local community members, the Wet Tropics being a good example. While most World Heritage management is about enforcing the protection and conservation of the area's outstanding universal value, just as important is the task of helping everyone appreciate why the site is significant. As some of the reasons for listing become increasingly technical, the need is stronger than ever to adequately communicate the special attributes of each site to locals and visitors.

The preamble to the World Heritage Convention notes threats to significant global heritage from various sources as a primary motivation for the development of the convention. Those threats, listed in the introduction to this book, are even more pressing today and additional threats have since emerged, notably the consequences of climate change and the impacts of greater global trade on biosecurity (the spread of invasive species). More than ever before, international cooperation for the protection of heritage is critical. Global climate change is a clear and recognised threat to most Australian sites, with the extreme weather of 2019 and disastrous fires in Tasmania adding further to concern. Other threats could follow from expanding industrial, mining, agricultural and tourism activities, as well as failed biosecurity and longer term economic disruption. These prospects add a great deal of uncertainty to the future and should encourage both a higher level of investment in protection and management of these remarkable places and also more vigorous identification and protection of all Australian heritage. The reality is that the approach to managing threats to World Heritage is largely dependent on the appetite of the government of the day to resist industry (and its associated financial and political incentives) in favour of conservation.

In the 2002 Budapest Declaration on World Heritage, the World Heritage Committee established several strategic objectives with a series of words starting with 'C': Credibility, Conservation, Capacity-building and Communication. The committee added a fifth 'C' in 2007 (Communities) and one might readily extend the series using the convention text. First, of course, is Cooperation—the essence of this convention and emphasised at the beginning of the Budapest Declaration. To this, the Budapest committee session added the following strategic objectives: Credibility, which points to the scientific and technical foundation for evaluation and monitoring; Conservation, which is the management needed to achieve protection; and Capacity-building, which is about the need for investment and sharing. To these overtly declared objectives, we can add engagement with Communities, which are the foundation of values and support for heritage; enhancing Communication so that we can better understand the global treasures on the list; and, finally, promoting Celebration, which takes us back to the wonders of these places and encourages us to enjoy what they are and what they stand for. One way in which Australia might contribute to these objectives is by better communicating the outstanding universal value of our World Heritage places so that more members of our society appreciate and support their long-term protection. I hope that this book will contribute in some way to that challenge.

World Heritage List Criteria

To be included on the World Heritage List, sites must be of outstanding universal value, meeting at least one of ten selection criteria:

i. represent a masterpiece of human creative genius

ii. exhibit an important interchange of human values, over a span of time or within a cultural area of the world, on developments in architecture or technology, monumental arts, town-planning or landscape design

iii. bear a unique or at least exceptional testimony to a cultural tradition or to a civilisation which is living or which has disappeared

iv. be an outstanding example of a type of building, architectural or technological ensemble or landscape which illustrates (a) significant stage(s) in human history

v. be an outstanding example of a traditional human settlement, land-use, or sea-use which is representative of a culture (or cultures), or human interaction with the environment especially when it has become vulnerable under the impact of irreversible change

vi. be directly or tangibly associated with events or living traditions, with ideas, or with beliefs, with artistic and literary works of outstanding universal significance. (The committee considers that this criterion should preferably be used in conjunction with other criteria)

vii. contain superlative natural phenomena or areas of exceptional natural beauty and aesthetic importance

viii. be outstanding examples representing major stages of earth's history, including the record of life, significant on-going geological processes in the development of landforms, or significant geomorphic or physiographic features

ix. be outstanding examples representing significant ongoing ecological and biological processes in the evolution and development of terrestrial, fresh water, coastal and marine ecosystems and communities of plants and animals

x. contain the most important and significant natural habitats for in-situ conservation of biological diversity, including those containing threatened species of outstanding universal value from the point of view of science or conservation

Criteria i–vi refer to cultural heritage and criteria vii–x refer to natural heritage. Sites can be listed under both cultural and natural criteria.

Responsibilities under the World Heritage Convention

The World Heritage Convention requires each signatory country to undertake a number of responsibilities:

Identification: to draw up a list of the cultural and natural heritage in its own territory and, where those properties are considered to be of outstanding universal value, to nominate them for World Heritage listing. This also means placing them on the national Tentative List, as required by the World Heritage Committee in advance of listing.

Protection: to establish a set of laws to formally protect its heritage. In Australia that might involve both state and federal legislation.

Conservation: to implement a formal management structure to ensure a World Heritage-listed property retains its outstanding universal value and that appropriate action is taken to minimise any threats. In Australia, this often involves state management agencies (such as national park management authorities or trusts, for example, the Sydney Opera House Trust); cultural heritage management authorities (including private or non-government organisations); and national organisations (such as the Great Barrier Reef Marine Park Authority).

Presentation: to develop and use various measures to help the community to understand the significance of a site and to interpret to visitors what the outstanding universal value of a site means. This frequently involves visitor centres, publications or other media and personal guides, often linked with tourism operations.

Transmission to future generations: to implement careful monitoring of the condition of the heritage to ensure that there is no decline in attributes or the integrity of each site so that the same outstanding universal value can be passed on to future generations. Regular reports on the condition of the site are part of this process.

Acknowledgements

As a young farmer near Bremer Bay, I began studies in biogeography at the University of Western Australia and later joined James Cook University, which supported my academic career. It is unlikely that I would have become involved with World Heritage were it not for three outstanding mentors. Professor Larry Hamilton invited me to the East-West Center in Hawaii, to supprt conservation across the Asia-Pacific region; he gave intellectual and practical support for many years. Bing Lucas was an exceptional chair of the IUCN World Commission on Protected Areas and recruited me to the commission. Dr Jim Thorsell, who was the IUCN natural heritage specialist for many years, helped me better understand the complexities of World Heritage.

I have been privileged to work with a large array of outstanding people in World Heritage practices, too many to mention all by name but I give thanks in particular to David Sheppard (IUCN), Prof. Adrian Phillips (United Kingdom), Dr Mechtild Rössler (World Heritage Centre), Dr Arturo Izurieta (Ecuador), Andrew Zaloumis (South Africa) and Australian colleagues Prof. Mike Archer AM, Dr Alastair Birtles, Prof. Ben Boer, David Cassells, Peter Cochrane, Dr Matt Curnock, Jon Day PSM, Dr David Gillieson, Allison Halliday, Dr Rosemary Hill, Peter Hitchcock AM, Prof. Marc Hockings, Dr Jacqueline Huggins AM, Graeme Kelleher AO, Helen Lardner, Prof. Carmen Lawrence, Bruce Leaver, Dr Jon Luly, Prof. Richard Mackay AM, Prof. Brendan Mackey, Andrew MacLean, Prof. Helene Marsh, Dr Dean Miller, Dr Warren Nicholls, Brian Prince, Phil Rist, Dr Lea Scherl, Joann Schmider, Dr Dermot Smyth, Dr Sharon Sullivan AO, Dr Leah Talbot, Howard Tanner AM, John Tanzer, Prof. Steve Turton, Dr Trevor Webb, Simon Woodley and Dr Graeme Worboys.

Particular thanks go to the management staff at Budj Bim for sharing their knowledge and supporting my learning about this site during my visit, especially Erin Rose, Aaron Morgan, Tyson Lovett-Murray and Sammy Walsh-Bannam. The many site managers who participated in the National World Heritage Forum in Canberra during November 2022 enhanced my knowledge and appreciation of our sites. I am grateful for the invitation to participate and for the support of Dr Sarah Tichen and Mia Thurgate, Department of Climate Change, Energy, the Environment and Water (DCCEEW), in facilitating valued discussions with site managers. I offer a special thank you to Dr Sue Sargent and the advisory committee for K'gari for helpful advice and comments. I also appreciate the support from Prof Richard Mackay on a number of cultural sites.

Many site managers provided advice and I am grateful to all, especially Ellen Weber, Tegan Gourley, Jason Irving, Angela Siggery, Lyn Wallace, Laura Matarese, Alida Eisermann, Michelle Stevenson, Jacqueline Reid, Dan Rosendahl, Cristina Venables, Elise Waugh, Rowena Hannaford, Alexa Shaw, Felicity Britten, Glenys Jones, Dianne Scopel and Natalie Lees. I thank my photographer friends Martin Willis, Steven Nowakowski, Angus Emmott and Nicolas Rakotopare, who shared some of their images and passion for these special places. Many people helped with information, especially Veronica Blazely, DCCEEW.

I am grateful to Penny Figgis AO, IUCN World Commission on Protected Areas Oceania Vice Chair, for encouraging me to write this book, and to the team at NLA Publishing for guiding me through the editing and publication process. The National Library is surely a magnificent resource for all Australians.

I particularly thank artist, wildlife photographer and friend Martin Willis, who suggested a major structural improvement. Polly Valentine, Leonie Valentine and Kate Spyby, my three daughters, made valuable suggestions. My wife Valerie was not only my first reader but shared much of the experience on which my knowledge has been based and managed to keep the home fires burning when I was away. Thank you all.

NLA Publishing would like to thank the following individuals and organisations for their assistance with FIrst Nations content: Anja Toms and Parks Australia (Kakadu) in joint management with Bininj/Mungguy Traditional Owners; Dan Rosendahl (Willandra Lakes); Uluru Media; Brad Moore and Shane Smith (Blue Mountains), and Gundungurra Traditional Owners.

References

I consulted a large number of sources in preparing this book, the most important of which were the original nomination documents for each site (prepared by Australia) and the corresponding evaluation reports by the IUCN and ICOMOS. Australia has a well-deserved reputation for the outstanding quality of its nomination documents, which include significant scientific and technical detail. I have also drawn upon the most recent operational guidelines prepared by the World Heritage Committee and on the text of the Convention itself. A particularly useful source is the Australian Committee of IUCN book published in 2013, *Keeping the Outstanding Exceptional: The Future of World Heritage in Australia*.

Sources for quoted material are listed in the order that they appear in this book. The pages on which the quotes appear are in bold. Sources in 'Other Sources' are listed alphabetically by author. Sources are cited in full in the chapter in which they first appear; in subsequent chapters the citations refer back to the first chapter in which they are mentioned.

GENERAL REFERENCE

Australia's National Heritage. Canberra: Australian Government Department of the Environment and Water Resources, 2007; Australian Heritage Council, *Australia's Fossil Heritage: A Catalogue of Important Australian Fossil Sites*. Collingwood, Vic.: CSIRO Publishing, 2012; Johnson, David, *The Geology of Australia*. Cambridge, Port Melbourne, Vic.: Cambridge University Press, 2004; 'Managing World Heritage in Australia', Australian Government Department of the Environment and Energy, environment.gov.au/heritage/about/world/management-australias-world-heritage-listed/managing-world-heritage-australia; Vickers-Rich, Patricia and Rich, Thomas Hewitt, *Wildlife of Gondwana*. Chatswood, NSW: Reed Books, 1993; White, Mary E., *The Greening of Gondwana*. Frenchs Forest, NSW: Reed, 1986.

INTRODUCTION TO WORLD HERITAGE

Sources of Quoted Material

1: 'Taj Mahal', UNESCO World Heritage Centre, whc.unesco.org/en/list/252; **3:** 'The World Heritage Convention', UNESCO World Heritage Centre, whc.unesco.org/en/convention; 'Convention Concerning the Protection of the World Cultural and Natural Heritage', UNESCO World Heritage Centre, 1972, whc.unesco.org/en/conventiontext; UNESCO Intergovernmental Committee for the Protection of the World Cultural and National Heritage, *Operational Guidelines for the Implementation of the World Heritage Convention*, Paris: World Heritage Centre, 2017, p.20, whc.unesco.org/en/guidelines.

Other Sources

'Arabian Oryx Sanctuary', UNESCO World Heritage Centre, whc.unesco.org/en/list/654; 'Cultural Landscape and Archaeological Remains of the Bamiyan Valley', UNESCO World Heritage Centre, whc.unesco.org/en/list/208; 'Dinosaur Provincial Park', UNESCO World Heritage Centre, whc.unesco.org/en/list/71; 'Dresden Elbe Valley', UNESCO World Heritage Centre, whc.unesco.org/en/list/1156; 'Messel Pit Fossil Site', UNESCO World Heritage Centre, whc.unesco.org/en/list/720; 'Willandra Lakes Region', UNESCO World Heritage Centre, whc.unesco.org/en/list/167.

KAKADU NATIONAL PARK

Breeden, Stanley and Wright, Belinda, *Kakadu: Looking After Country the Gagudju Way*. Brookvale, NSW: Simon & Schuster, 1989; *Commonwealth Aboriginal Land Rights (Northern Territory) Act 1976* (Cth); International Council on Monuments and Sites, 'ICOMOS Evaluation: The National Park of Kakadu', World Heritage Centre, 1980, whc.unesco.org/en/list/147/documents; International Union for Conservation of Nature, *Kakadu National Park—2020 Conservation Outlook Assessment*, worldheritageoutlook.iucn.org/; International Union for Conservation of Nature, 'World Heritage Nomination IUCN Technical Review: Kakadu National Park', World Heritage Centre, 1981, whc.unesco.org/en/list/147/documents; 'Kakadu National Park', UNESCO World Heritage Centre, whc.unesco.org/en/list/147; *Kakadu National Park Visitor Guide and Maps*. Jabiru, NT: Kakadu National Park Board of Management, 1997; Meier, Leo and Balderstone, Simon, *Kakadu: A Heritage for the Future*. McMahons Point: Weldons, 1987; Morris, Ian, *Kakadu National Park, Australia. Natural History Guide*. Brisbane: Steve Parish, Brisbane, 1996; Neidjie, Bill, cited in S. Breedon and B. Wright, p.24. See ref in Kakadu National Park ch; *Nomination of Kakadu National Park for Inclusion in the World Heritage List*. Canberra: Australian National Parks and Wildlife Service, 1980, © Commonwealth of Australia; Ovington, Derrick, *Kakadu: A World Heritage of Unsurpassed Beauty*. Canberra: Australian Government Publishing Service, 1986.

GREAT BARRIER REEF

Carter, E. and Thulstrup, H., *Report on the Joint World Heritage Centre/IUCN Reactive Monitoring Mission to the Great Barrier Reef (Australia) from 21 to 30 March 2022*, World Heritage Centre, 2022, whc.unesco.org/en/list/154/documents/; Commonwealth of Australia, *State Party Report on the State of Conservation for Australia's Great Barrier Reef*, World Heritage Centre, 2022, whc.unesco.org/en/list/154/documents/; 'Great Barrier Reef', UNESCO World Heritage Centre, whc.unesco.org/en/list/154; *Great Barrier Reef Marine Park Act 1975* (Cth); Great Barrier Reef Marine Park Authority, *Great Barrier Reef Outlook Report 2014*. Townsville, Qld: Great Barrier Reef Marine Park Authority, 2014; International Union for Conservation of Nature, *Great Barrier Reef—2020 Conservation Outlook Assessment*, worldheritageoutlook.iucn.org/; International Union for Conservation of Nature, 'World Heritage Nomination IUCN Technical Review: The Great Barrier Reef', World Heritage Centre, 1981, whc.unesco.org/en/list/154/documents; Lucas, P.H.C., Webb, T., Valentine, P.S. and Marsh, H., *The Outstanding Universal Value of the Great Barrier Reef World Heritage Area*. Townsville, Qld: Great Barrier Reef Marine Park Authority, 1997; McCalman, Iain, *The Reef: A Passionate History*. Melbourne: Penguin Random House, 2016; *Nomination of the Great Barrier Reef by the Commonwealth of Australia for Inclusion in the World Heritage List*. Townsville, Qld: Great Barrier Reef Marine Park Authority, 1981, © Commonwealth of Australia; Thorsborne, Margaret and Thorsborne, Arthur, *Hinchinbrook Island: The Land Time Forgot*. McMahons Point, NSW: Weldons, 1988; Veron, J.E.N., *A Reef in Time: The Great Barrier Reef from Beginning to End*. Cambridge, Mass.: Harvard University Press, 2008; Wright, Judith, *The Coral Battleground*. West Melbourne, Vic.: Thomas Nelson Australia, 1977.

WILLANDRA LAKES REGION

International Council on Monuments and Sites, 'ICOMOS Evaluation: The Willandra Lakes Region', World Heritage Centre, 1981, whc.unesco.org/en/list/167/documents; International Union for Conservation of Nature, 'World Heritage Nomination IUCN Technical Review: Willandra Lakes Region', World Heritage Centre, 1981, whc.unesco.org/en/list/167/documents; Johnston, Harvey and Mintern, Richard, 'Managing Australia's World Heritage in the Willandra Lakes Region', in Penny Figgis et al. (eds), *Keeping the Outstanding Exceptional: The Future of World Heritage in*

Australia. Sydney: Australian Committee for IUCN, 2013, pp.102–107; *Nomination of the Willandra Lakes Region for Inclusion in the World Heritage List*. Canberra: Australian Heritage Commission, 1981, © Commonwealth of Australia; 'Willandra Lakes Region', UNESCO World Heritage Centre, whc.unesco.org/en/list/167.
in Willandra Lakes Region ch.

LORD HOWE ISLAND GROUP

Environment Protection and Biodiversity Conservation Act 1999 (Cth); Frith, Clifford B., *The Woodhen: A Flightless Island Bird Defying Extinction*. Collingwood, Vic.: CSIRO Publishing, 2013; International Union for Conservation of Nature, 'World Heritage Nomination IUCN Technical Review: Lord Howe Island Group', World Heritage Centre, 1982, whc.unesco.org/en/list/186/documents; 'Lord Howe Island Group', World Heritage Centre, whc.unesco.org/en/list/186; MacArthur, Robert H. and Wilson, Edward O., *The Theory of Island Biogeography*. Princeton, N.J.: Princeton University Press, 1967; New South Wales Government et al, *Nomination of the Lord Howe Island Group by the Commonwealth of Australia for Inclusion on the World Heritage List*. Canberra: Australian Heritage Commission, 1981, © Commonwealth of Australia; Wills, Stephen, 'Engaging the Community as Volunteers—The Case of Lord Howe Island World Heritage Area' in Penny Figgis et al. (eds), pp.102–107. See ref in Willandra Lakes Region ch.

TASMANIAN WILDERNESS

Angus, Max, *The World of Olegas Truchanas*. Hobart: Olegas Truchanas Publication Committee, 1975; Bowdler, Sandra, 'The Empty Coast: Conditions for Human Occupation in Southeast Australia during the Late Pleistocene' in Haberle et al. (eds), *Altered Ecologies: Fire, Climate and Human Influence on Terrestrial Landscapes*. Canberra: ANU E Press, 2010, pp.177–185; Brown, Bob and Kiernan, Kevin, *Lake Pedder*. Hobart: Wilderness Society, 1986; Commonwealth Department of the Arts, Sport, the Environment, Tourism and Territories and Government of the State of Tasmania, *Nomination of the Tasmanian Wilderness by the Government of Australia for Inclusion in the World Heritage List*. Canberra: Department of Arts, Sport, the Environment, Tourism and Territories, 1988, © Commonwealth of Australia; Commonwealth of Australia, *State Party Report on the State of Conservation of the Tasmanian Wilderness*, World Heritage Centre, 2022/2023, whc.unesco.org/en/list/181/documents/; *Commonwealth v Tasmania* (1983) 158 CLR 1; Hungerford, Alice, *Upriver: Untold Stories of the Franklin River Activists*. Maleny, Qld: UpRiver Mob, 2013; International Council on Monuments and Sites, 'ICOMOS Evaluation: Western Tasmania Wilderness National Parks', World Heritage Centre, 1981, whc.unesco.org/en/list/181/documents; International Union for Conservation of Nature, 'World Heritage Nomination IUCN Technical Review: Western Tasmania Wilderness National Parks', World Heritage Centre, 1982, whc.unesco.org/en/list/181/documents; 'Port Davey Marine Reserve', Tasmania Parks & Wildlife Service, parks.tas.gov.au/index.aspx?base=3126; Tasmanian Government and Australian Heritage Commission, *Nomination of Western Tasmania Wilderness National Parks by the Commonwealth of Australia for Inclusion in the World Heritage List*. Canberra: Australian Heritage Commission, 1981, © Commonwealth of Australia; 'Tasmanian Wilderness', World Heritage Centre, whc.unesco.org/en/list/181.

GONDWANA RAINFORESTS OF AUSTRALIA

Adam, Paul, *New South Wales Rainforests: The Nomination for the World Heritage List*. Sydney: National Parks and Wildlife Service of New South Wales, 1987, © Commonwealth of Australia; Adam, Paul, *Australian Rainforests*. Oxford: Oxford University Press, 1992; 'Canadian Rocky Mountain Parks', World Heritage Centre, whc.unesco.org/en/list/304; Commonwealth of Australia, *Gondwana Rainforests of Australia State of Conservation Update—April 2020*, unesco.org/en/soc/4173; Department of Climate Change, Energy, the Environment and Water, *State Party Report on the State of Conservation of the Gondwana Rainforests of Australia (Australia)*, World Heritage Centre, 2022, whc.unesco.org/en/list/368/documents/; 'Gondwana Rainforests of Australia', World Heritage Centre, whc.unesco.org/en/list/368; Hunter, R. John, *World Heritage and Associative Natural Values of the Central Eastern Rainforest Reserves of Australia*. Hurstville, NSW: NSW National Parks and Wildlife Service, 2003; International Union for Conservation of Nature, 'World Heritage Nomination IUCN Summary: Central Eastern Rainforests of Australia (Extension)', World Heritage Centre, 1993, whc.unesco.org/en/list/368/documents; International Union for Conservation of Nature, 'World Heritage Nomination IUCN Summary: New South Wales Rain Forests', World Heritage Centre, 1986, whc.unesco.org/en/list/368/documents Waters, Tricia, 'Gondwana Rainforests of Australia World Heritage Area—Remnants of Our Primeval Past' in Penny Figgis et al. (eds), pp.108–111. See ref in Willandra Lakes Region ch.

ULUṞU-KATA TJUṮA NATIONAL PARK

Sources of Quoted Material

97: 'Tjukurpa', Australian Institute of Aboriginal and Torres Strait Islander Studies, aiatsis.gov.au/exhibitions/Tjukurpa.

Other Sources

Breeden, Stanley, *Uluru: Looking after Uluru-Kata Tjuta the Anangu Way*. Marleston, SA: J.B. Books, 1997; International Council on Monuments and Sites, 'Uluru', World Heritage Centre, 1994, whc.unesco.org/en/list/447/documents; International Union for Conservation of Nature, 'World Heritage Nomination IUCN Summary: Uluru (Ayres Rock–Mount Olga) National Park (Australia)', World Heritage Centre, 1987, whc.unesco.org/en/list/447/documents; *Nomination of Uluru (Ayers Rock-Mount Olga) National Park for Inclusion on the World Heritage List*. Canberra: Australian National Parks and Wildlife Service, 1986, © Commonwealth of Australia; *Renomination of Uluru-Kata Tjuta National Park by the Government of Australia for Inscription on the World Heritage List*. Canberra: Department of the Environment, Sport and Territories, 1994, © Commonwealth of Australia; 'The People and the Place: Uluru–Kata Tjuta National Park', Parks Australia, parksaustralia.gov.au/uluru/people-and-place.html.

WET TROPICS OF QUEENSLAND

Sources of Quoted Material

125: International Union for Conservation of Nature, 'World Heritage Nomination IUCN Summary: Wet Tropical Rainforests (North-east Australia)', World Heritage Centre, 1988, whc.unesco.org/en/list/186/documents.

Other Sources

Breeden, Stanley, *Visions of a Rainforest: A Year in Australia's Tropical Rainforest*. East Roseville, NSW: Simon & Schuster Australia, 1992; Clarke, Campbell (ed.), *From the Heart: Celebrating 20 Years of the Wet Tropics of Queensland World Heritage Area*. Cairns, Qld: Wet Tropics Management Authority, 2009; International Union for Conservation of Nature, *Wet Tropics of Queensland—2020 Conservation Outlook Assessment*, worldheritageoutlook.iucn.org/; International Union for Conservation of Nature, 'World Heritage Nomination IUCN Summary: Wet Tropical Rainforests (North-east Australia)', World Heritage Centre, 1988, whc.unesco.org/en/list/578/documents; *Nomination of Wet Tropical Rainforests of North-east Australia by the Government of Australia for Inclusion in the World Heritage List*. Canberra: Department of the Arts, Sport, the Environment, Tourism and Territories, 1987, © Commonwealth of Australia; Russell, Rupert, *Daintree: Where the Rainforest Meets the Reef*. McMahons Point, NSW: Kevin Weldon and Associates; Melbourne: Australian Conservation Foundation, 1985; Stork, Nigel and Turton, Stephen M. (Eds), *Living in a Dynamic Tropical Forest Landscape*. Malden,

MA: Blackwell Publishers, 2008; Trott, Laurie, *Wet Tropics in Profile: A Reference Guide to the Wet Tropics of Queensland World Heritage Area*. Cairns, Qld: Wet Tropics Management Authority, 1996; 'Wet Tropics of Queensland', UNESCO World Heritage Centre, whc.unesco.org/en/list/486; Wet Tropics Management Authority, *State of the Wet Tropics 2020–2021*, 2021, wettropics.gov.au/site/user-assets/docs/SoWT/2020-2021%20State%20of%20Wet%20Tropics.pdf; Wet Tropics Management Authority, *State of the Wet Tropics 2021–2022*, 2022, wettrop-ics.gov.au/site/user-assets/docs/2021-22%20State%20of%20Wet%20Tropics%20report.pdf; Wilkie, Bill, *The Daintree Blockade: The Battle for Australia's Tropical Rainforests*. Mossman, Qld: Fourmile Books, 2017.

SHARK BAY, WESTERN AUSTRALIA

International Union for Conservation of Nature, 'World Heritage Nomination IUCN Summary: Shark Bay (Australia)', World Heritage Centre, 1991, whc.unesco.org/en/list/578/documents; Mann, Jess, 'World Heritage in Western Australia' in Figgis et al., pp.88–95. See ref in Willandra Lakes Region ch; McCluskey, Paul, *Shark Bay World Heritage Property: Strategic Plan 2008–2020*. Perth: Department of Environment and Conservation, 2008, © Commonwealth of Australia; *Nomination of Shark Bay, Western Australia by the Government of Australia for Inclusion in the World Heritage List*. Canberra: Department of the Arts, Sport, the Environment, Tourism and Territories, 1990, © Commonwealth of Australia; 'Shark Bay', UNESCO World Heritage Centre, whc.unesco.org/en/list/578; *Shark Bay Region Plan*. Perth: Western Australia State Planning Commission and Western Australia Department of Conservation and Land Management, 1987.

K'GARI (FRASER ISLAND)

Butchulla Aboriginal Corporation RNTBC, *Butchulla Aboriginal Corporation Strategic Business Plan 2020-2030*, 2020, bac-kgari.org/media/5uohttwb/bac-strategic-business-plan-2020.pdf; 'Fraser Island', UNESCO World Heritage Centre, whc.unesco.org/en/list/630; Hemmings, L. and Sinclair, J, *Nomination of the Great Sandy Region for Inclusion in the World Heritage List*. Melbourne: Australian Conservation Foundation, 1984, © Commonwealth of Australia; International Union for Conservation of Nature, 'World Heritage Nomination IUCN Summary: Fraser Island and the Great Sandy Region (Australia)', World Heritage Centre, 1992, whc.unesco.org/en/list/630/documents; Miles, Steven, 'National Park Renaming Recognises Traditional Owners', The Queensland Cabinet and Ministerial Directory, statements.queensland.gov.au/statement/2017/4/24/national-park-renaming-recognises-traditional-owners; Sinclair, John, *Fraser Island and Cooloola*. Sydney: Weldon, 1990; Stringer, Angie, 'Managing Fraser Island World Heritage Area' in Penny Figgis et al. (eds), pp.82–87. See ref in ch 3.

AUSTRALIAN FOSSIL MAMMAL SITES (RIVERSLEIGH/NARACOORTE)

Archer, Michael et al., *Riversleigh: The Story of Animals in Ancient Rainforests of Inland Australia*. Balgowlah, NSW: Reed Books, 1991; 'Australian Fossil Mammal Sites (Riversleigh/Naracoorte)', World Heritage Centre, whc.unesco.org/en/list/698; Craven-Carden, Deborah and Stringer, Angie, 'The Australian Fossil Mammal Sites: Naracoorte and Riversleigh' in Penny Figgis et al. (eds), pp.112–117. See ref ch 3; International Union for Conservation of Nature, 'World Heritage Nomination IUCN Summary: Australian Fossil Sites', World Heritage Centre, 1994, whc.unesco.org/en/list/698/documents; Luly, J. and Valentine, P.S., *On the Outstanding Universal Value of the Australian Fossil Mammal Site (Riversleigh/Naracoorte) World Heritage Area: A Report to the World Heritage Unit, DEST*. Townsville, Qld: School of Tropical Environmental Studies and Geography, James Cook University, 1998; *The Origin and Evolution of Australia's Mammals: Nomination of Australian Fossil Mammal Sites (A Serial Nomination of Sites at Murgon, Riversleigh and Naracoorte) by the Government of Australia for Inscription on the World Heritage List*. Canberra: Department of the Environment, Sport and Territories, 1993, © Commonwealth of Australia; Wells, R.T. et al., 'The Fossil Vertebrate Deposits of Victoria Fossil Cave Naracoorte: An Introduction to the Geology and Fauna', *The Australian Zoologist*, vol.21, 1984, pp.305–333.

MACQUARIE ISLAND

Sources of Quoted Material

171: 'Macquarie Island', World Heritage Centre, whc.unesco.org/en/list/629.

Other Sources

Edge of Nowhere [videorecording], dir. Jeremy Hogarth, Screen Queensland, 2011; International Union for Conservation of Nature, 'World Heritage Nomination IUCN Technical Evaluation: Macquarie Island (Australia)', World Heritage Centre, 1997, whc.unesco.org/en/list/629/documents; *Nomination for World Heritage Listing: Macquarie Island Nature Reserve*. Hobart: Tasmanian Department of Parks, Wildlife and Heritage, 1990, © Commonwealth of Australia; Salton, M. et al., 'Population Status of the Endemic Royal Penguin (Eudyptes schlegeli) at Macquarie Island', *Polar Biology*, 2019, vol.42, pp.771–781, doi.org/10.1007/s00300-019-02470-y; 'World Heritage Places—Macquarie Island', Australian Government Department of the Environment and Energy, environment.gov.au/heritage/places/world/macquarie-island.

HEARD AND MCDONALD ISLANDS

'Heard and McDonald Islands', World Heritage Centre, whc.unesco.org/en/list/577; *Heard Island and McDonald Islands: Nomination by the Government of Australia for Inscription on the World Heritage List 1996*. Canberra: Department of the Environment, Sport & Territories, 1996, © Commonwealth of Australia; International Union for Conservation of Nature, 'World Heritage Nomination IUCN Technical Evaluation: Heard Island and McDonald Islands (Australia)', World Heritage Centre, 1997, whc.unesco.org/en/list/577/documents; International Union for Conservation of Nature Commission on National Parks and Protected Areas, *The World's Greatest Natural Areas: An Indicative Inventory of Natural Sites of World Heritage Quality*. Gland, Switzerland: International Union for Conservation of Nature, 1982; Keage, Peter L., 'The Conservation Status of Heard Island and the McDonald Islands', Masters thesis, University of Tasmania, 1981, eprints.utas.edu.au/20090/1/whole_KeagePeterLawson1981_thesis.pdf.

GREATER BLUE MOUNTAINS AREA

Sources of Quoted Material

197: Darwin, Charles, *Voyage of the Beagle: Charles Darwin's Journal of Researches*. London: Penguin Books, 1989, pp.319–320.

Other Sources

Australia, *The Greater Blue Mountains Area: World Heritage Nomination*. Canberra: NSW Parks and Wildlife Service and Environment Australia, 1998; Commonwealth of Australia, *State of Conservation Greater Blue Mountains Area*, World Heritage Centre, 2021, whc.unesco.org/en/soc/4174; Department of Climate Change, Energy, the Environment and Water, *State Party Report on the State of Conservation of the Greater Blue Mountains Area (Australia)*, World Heritage Cen-tre, 2022, whc.unesco.org/en/list/917/documents/; *Environment Protection and Biodiversity Conservation Act 1999* (Cth); 'Greater Blue Mountains Area', World Heritage Centre, whc.unesco.org/en/list/917; International Council on Monuments and Sites, 'Greater Blue Mountains (Australia)', World Heritage Centre, 1998, whc.unesco.org/en/list/917/documents; International Union for Conservation of Nature, *Greater Blue Mountains Area—2020 Conservation Outlook Assessment*, worldheritageoutlook.iucn.org/; International Union for Conservation of Nature, 'World Heritage Nomination IUCN Technical Evaluation: The Greater Blue Mountains Area (Australia)', World Heritage Centre, 1999, whc.unesco.org/en/list/917/documents; Reid, Jacqueline, 'Managing Australia's World Heritage in the Greater Blue Mountains' in Penny Figgis, et al. (eds), pp.96–101. See ref in ch 3; Smith, Judy and Smith, Peter, 'Native Fauna of the Greater Blue Mountains World Heritage Area', Blue Mountains Nature, 2017, bmnature.info/fauna-gbmwha.shtml; Smith, P., *Impact of the 2019–20 Fires on the Greater Blue Mountains World Heritage Area—Version 2: Report to Blue Mountains Conservation Society*, 2021, 10.13140/RG.2.2.26138.18881/2.

PURNULULU NATIONAL PARK

Sources of Quoted Material

217: Australia, *Nomination of Purnululu National Park by the Government of Australia for Inscription on the World Heritage List*. Canberra: Environment Australia, 2002, p.20, © Commonwealth of Australia.

Other Sources

International Council on Monuments and Sites, 'Purnululu National Park (Australia)', World Heritage Centre, 2003, whc.unesco.org/en/list/1094/documents; International Union for Conservation of Nature, 'World Heritage Nomination IUCN Technical Evaluation: Purnululu National Park (Australia)', World Heritage Centre, 2003, whc.unesco.org/en/list/1094/documents; Mann, Jess. See ref in ch 9; Olsen, John, *The Land Beyond Time: A Modern Exploration of Australia's North-west Frontiers*. South Melbourne: Macmillan, 1984; 'Purnululu National Park', World Heritage Centre, whc.unesco.org/en/list/1094; *The Wonder of Western Australia* [videorecording], dir. Guy Baskin, Perth: Swan Television and Radio Broadcasters, between 1982 and 1984.

ROYAL EXHIBITION BUILDING AND CARLTON GARDENS

International Council on Monuments and Sites, 'Royal Exhibition Building (Australia)', World Heritage Centre, 2004, whc.unesco.org/en/list/1131/documents; *Nomination of Royal Exhibition Building and Carlton Gardens, Melbourne, by the Government of Australia for Inscription on the World Heritage List*. Canberra: Environment Australia, 2002, © Commonwealth of Australia; *Melbourne International Exhibition 1880–1881: Official Record*. Melbourne: Mason, Firth & M'Cutcheon, 1882, handle.slv.vic.gov.au/10381/298884; 'Royal Exhibition Building and Carlton Gardens', World Heritage Centre, whc.unesco.org/en/list/1131.

SYDNEY OPERA HOUSE

Sources of Quoted Material

233: *Sydney Opera House Nomination for World Heritage Listing: Executive Summary*. World Heritage Centre, whc.unesco.org/uploads/nominations/166rev.pdf, p.ii, © Commonwealth of Australia; *Sydney Opera House: Nomination by the Government of Australia for Inscription on the World Heritage List 2006*. Canberra: Department of the Environment and Heritage, 2006, p.22 © Commonwealth of Australia; **237:** *Nomination by the Government of Australia*, pp.44, 16, 13, 31; **241:** International Council on Monuments and Sites, 'Sydney Opera House (Australia)', World Heritage Centre, 2007, whc.unesco.org/en/list/166/documents, p.89; World Heritage Committee, 'Report of the Rapporteur: Bureau of the World Heritage Committee Fifth Session', World Heritage Centre, 1981, whc.unesco.org/archive/repbur81.htm; International Council on Monuments and Sites, 'Sydney Opera House (Australia)', World Heritage Centre, 2007, whc.unesco.org/en/list/166/documents, p.92.

Other Sources

'Badu Gili', Sydney Opera House, sydneyoperahouse.com/visit-us/BaduGili.html; Croker, Alan, *Respecting the Vision: Sydney Opera House —A Conservation Management Plan*, 4th edn. Sydney: Sydney Opera House, 2017; 'The National: Art Gallery of NSW', This Is No Fantasy, thisisnofantasy.com/exhibition/the-national-new-australian-art; 'Sydney Opera House', World Heritage Centre, whc.unesco.org/en/list/166; Sydney Opera House Trust, *Annual Report Sydney Opera House 2021–2022*, 2021.

AUSTRALIAN CONVICT SITES

'Australian Convict Sites', World Heritage Centre, whc.unesco.org/en/list/1306; *Australian Convict Sites: World Heritage Nomination*. Canberra: Department of the Environment, Water, Heritage and the Arts, 2008, p.72, © Commonwealth of Australia; Clarke, Marcus, *For the Term of His Natural Life*. London: The UK Bureau, 2017, available from: Amazon Kindle; Hughes, Robert, *The Fatal Shore*. New York: Knopf, 1986; International Council on Monuments and Sites, 'Australian Convict Sites (Australia)', World Heritage Centre, 2010, whc.unesco.org/en/list/1306/documents; *Medical Journal of the John Barry, Male Convict Ship, 1821*, The National Archives (UK), ADM 101/28/2, nationalarchives.gov.uk/documents/adm101-38-2.pdf; 'Surgeons at Sea: Royal Navy Medical Officers' Journals', The National Archives (UK), nationalarchives.gov.uk/surgeonsatsea.

NINGALOO COAST

Sources of Quoted Material

261: 'Ningaloo Coast', World Heritage Centre, whc.unesco.org/en/list/1369.

Other Sources

International Union for Conservation of Nature, *Ningaloo Coast—2020 Conservation Outlook Assessment*, worldheritageoutlook.iucn.org/; International Union for Conservation of Nature, 'World Heritage Nomination IUCN Technical Evaluation: Ningaloo Coast (Australia)', World Heritage Centre, 2011, whc.unesco.org/en/list/1369/documents; Mann, Jess. See ref in ch 9; *Ningaloo Coast: World Heritage Nomination*. Canberra: Department of the Environment, Water, Heritage and the Arts 2010, © Commonwealth of Australia.

BUDJ BIM CULTURAL LANDSCAPE

'Budj Bim Cultural Landscape', UNESCO World Heritage Centre, whc.unesco.org/en/list/1577; Department of the Environment and Energy, *Budj Bim Cultural Landscape: World Heritage Nomination*. World Heritage Centre, 2017, whc.unesco.org/en/list/1577/documents/; Gunditjmarra People and Wettenhall, Gib, *The People of Budj Bim*. Heywood: emPress, 2022 (2nd edition); Matchan, E.L., Phillips, D., Jourdan, F. and Oostingh, K., 'Early Human Occupation of Southeastern Australia: New Insights from 40Ar/39Ar Dating of Young Volcanoes', *Geology*, vol.48, p.390–394, doi.org/10.1130/G47166.1.

POTENTIAL NEW WORLD HERITAGE SITES IN AUSTRALIA

Figgis, Penny et al. (eds), *Keeping the Outstanding Exceptional: The Future of World Heritage in Australia*. Sydney: IUCN Australian Committee of IUCN, 2013.

List of Illustrations

All images are from National Library of Australia collections, except where otherwise noted. Page numbers in this book are given in bold type. To view catalogue information and full National Library images online, enter into your browser **nla.gov.au/** followed by the item's catalogue identifier (e.g. to see Peter Dombrovskis' photograph of native pigface in the Tarkine wilderness on page vi, enter nla.gov.au/nla.cat-vn682098).

Peter Dombrovskis images appear courtesy of Liz Dombrovskis.
Peter Valentine images appear courtesy of Peter Valentine.
Leonie Valentine images appear courtesy of Leonie Valentine.
Shutterstock images are used under licence from Shutterstock.com.
A number of images have been reproduced according to various licences. These licences have been abbreviated in the captions. To view the relevant licence online, refer to the following list:
CC0 1.0: creativecommons.org/publicdomain/zero/1.0
CC BY 2.0: creativecommons.org/licenses/by/2.0
CC BY-ND 2.0: creativecommons.org/licenses/by-nd/2.0
CC BY-SA 2.0: creativecommons.org/licenses/by-sa/2.0

FRONT COVER

Peter Dombrovskis (1945–1996), *Mount Geryon from the Labyrinth, Du Cane Range, Tasmania*, 1986, nla.cat-vn4900025.

SPINE

Tororo Reaction, *Aerial View of Great Barrier Reef with Clear Blue Water, Queensland, Australia*, Shutterstock, 17708278.

BACK COVER

Nick Fox, *Sunset Over the Famous Walls of China in Mungo National Park, New South Wales, Australia*, Shutterstock, 696080992.

ENDPAPERS

Courtesy Mike Ellott.

PRELIMINARY PAGES

ii–iii Cornfield, *Sunset at a Desert in the Mungo National Park, New South Wales, Australia*, Shutterstock, 629082077; **v** National Library of Australia, *Bob Brown*, 2017; **vi** Peter Dombrovskis (1945–1996), *Native Pigface, Tarkine Wilderness, Tasmania*, 1995, nla.cat-vn682098.

INTRODUCTION

2 Nick Rains, *Uluru Kata Tjuta National Park, NT*, 1995, Tourism Australia, 607893; **4** Doleesi, *Statues in front of Abu Simbel Temple in Aswan Egypt*, Shutterstock, 1076006336.

KAKADU

6–7 Parks Australia, *Barramundi, Nanguluwurr Rock Art Site, Kakadu*, 2014, courtesy Parks Australia; **8–9** John Paulsen, *Panorama 41, Kakadu, Australia*, 2013, flickr.com/photos/121422802@N04/13414201993, reproduced under CC BY 2.0; **10 (top)** Parks Australia, *Hand Stencils, Nanguluwur Rock Art Site, Kakadu*, 2014, courtesy Parks Australia; **(bottom)** Parks Australia, *Ship, Nanguluwur Rock Art Site, Kakadu*, 2014, courtesy Parks Australia; **12** Janelle Lugge, *Aerial View of Jim Jim Waterfall, Kakadu*, Shutterstock, 1482842129; **13** Sarah George (b.1966), *Big Bill Neidjie Contemplates the Impact of Uranium Mining on Djabulugka Billabong, Jabiluka, Kakadu National Park*, 1995, nla.cat-vn4367339; **15** Simon O'Dwyer (b.1969), *Magpie Geese in the Wetlands Which Are Slightly Above Sea Level and at Risk of Salt Water Contamination, Kakadu National Park, Northern Territory, 10 November 2005*, nla.cat-vn4533111; **16** R. Weisswalk, *Large, Adult Male Saltwater Crocodile in the Shallow Water of Yellow Water-Ngurrungurrudjba Billabong*, Shutterstock, 2211250197; **18–19** 2021 Photography, *Dusk at Ubirr Rock Looking down at the Sun Reflected in the Waters of the Nadab Floodplains*, Shutterstock, 617365613; **20** Allan Fox (1931–2013, photographer), Australian National Parks and Wildlife Service, *Outliers, Kakadu National Park*, c.1980, nla.cat-vn5810186, courtesy Estate of Allan Fox; **21** Lyn McLeavy (b.1948), *Land Rights Not Uranium, Independence for Australia!, MAUM Rally, Melbourne*, 1979, nla.cat-vn40398; **22** Richard Woldendorp (b.1927), *Controlled Burning, Kakadu National Park, Northern Territory*, 1998, nla.cat-vn3102316; **23** Simon O'Dwyer (b.1969), *Elder Violet Lawson in the Freshwater Wetlands with Lotus Lilies Which Are at Risk of Salt Water Contamination Due to Rising Sea Levels, Kakadu National Park, Northern Territory, 10 November 2005*, nla.cat-vn4533118

GREAT BARRIER REEF

24–25 Michael Smith ITWP, *A Turtle Diving Back to the Reef in a Shallow Lagoon on Lady Elliot Island*, Shutterstock, 549545044; **26** Richard Woldendorp (b.1927), *Barrier Reef, off Gladstone, Queensland*, 1996, nla.cat-vn3102383; **27** imageBroker, *Dugong (Dugong dugon) Feeding on Seagrass Bed, with Golden Trevallys (Gnathanodon speciosus), Great Barrier Reef*, Alamy Stock Photo, CRA6XN; **28–29** Darren Jew, Tourism and Events Queensland, *Pixies Garden, Ribbon Reef, Great Barrier Reef, QLD* (detail), 2010, Tourism Australia, 641033; **30** Brian Gratwicke, *Blue-Green Chromis*, 2011, flickr.com/photos/briangratwicke/5421442899, reproduced under CC BY 2.0; **32–33** Matt Curnock, *Dwarf Minke Whale*, 2018, courtesy Matt Curnock; **34** Coral Brunner, *Acropora Coral Spawning on Magnetic Island*, Shutterstock, 1556924162; **35** A. Elliott, *Acropora formosa Showing Individual Eggs Leaving Polyps*, 1981, Commonwealth of Australia (GBRMPA), 108637; **36** The Edit Suite, *Acquscene Magnetic Island*, Tourism and Events Queensland; **37** D. Wachenfeld, *A Marine Biologist Injects a Crown of Thorns Starfish Feeding on an Acropora Plate Coral, with CuSO4*, 1999, Commonwealth of Australia (GBRMPA), 115423; **38** Paul Marshall, *Bleached Coral in the Waters of Keppel Island*, Commonwealth of Australia (GBRMPA), 134053; **39** Bob Nicol, Australian Information Service, *Biologist Dr Robert Endean Examining a Crown of Thorns Starfish in His Laboratory at the University of Queensland, 2 July 1970*, nla.cat-vn7761872.

WILLANDRA LAKES REGION

40–41 Leah-Anne Thompson, *Beautiful Scenery in Outback Australia, Willandra Lakes World Heritage Area*, Shutterstock, 1091830295; **42–43** Wolfgang Sievers (1913–2007), Plate two in *The Mystery of the Mungo Man, New South Wales*, 1994, nla.cat-vn993430; **44** Cornfield, *Sunset at a Desert in the Mungo National Park, New South Wales, Australia*, Shutterstock, 629082077; **45** Michael Amendolia, *Pleistocene Human Footprints, Willandra Lakes, New South Wales*, 2005, courtesy Michael Amendolia; **46–47** Leah-Ann Thompson, *Outback Landscape with Delicate Structures at Red Top Along the Large Lunette Formed by Wind and Water Erosion Along A Dried Up Lake*, Shutterstock, 794485033; **48** Ian Evans, *Kangaroo at Mungo National Park NSW*, Shutterstock, 1142336069; **49** Michael Amendolia, *Elders Representing the Three Traditional Groups of the Willandra Lakes World Heritage Area, the Barindji, the Mutthi Mutthi and Ngiyampaa, Look at What is Considered to be the Largest Collection of Pleistocene Human Footprints in Australia*, 2005, courtesy Michael Amendolia.

LORD HOWE ISLAND

50–51 Chris Howarth, *Lord Howe Island Sunset*, Alamy Stock Photo, DJ7K55; **52** Tony Feder, *Banyan Tree at Lord Howe Island*, iStock, 1324989516; **53** John Carnemolla, *Lord Howe Island*, Shutterstock, 521611984; **54** Pauline and John Grimshaw, *Lord Howe Island, Ball's Pyramid, Observatory Rock and Wheatsheaf Islet*, 2014, flickr.com/photos/30593522@N05/15524240899, courtesy Pauline and John Grimshaw; **55** Leonie Valentine, *Providence Petrel*, 2018; **56** William Thomas Cooper (1834–2015), *Orange-fronted Parakeet (Cyanoramphus malherbi), Red-fronted*

Dombrovskis (1945–1996), *Satinay Forest, Fraser Island, Queensland, 1995*, nla.cat-vn3915728; **144** Peter Valentine, *Ground Parrot, Fraser Island*, 2016; **146–147** Peter Dombrovskis (1945–1996), *Mangroves at Sunset, Fraser Island, Queensland*, 1995, nla.cat-vn4803799; **148** Darren Tierney, *Dingo on Fraser Island*, Shutterstock, 501909724; **149** courtesy Steven Nowakowski; **150** courtesy Steven Nowakowski; **151 (both)** courtesy Butchulla Aboriginal Corporation.

AUSTRALIAN FOSSIL MAMMAL SITES (RIVERSLEIGH/NARACOORTE)

152–153 Peter Schouten, *Reconstruction of Microleo attenboroughi*, c.2016, courtesy Peter Schouten; **154 (main)** *Riversleigh Overview*, c.2010, Australian Heritage Photographic Library, Department of the Environment and Energy, rt56496, courtesy Department of the Environment and Energy; **(inset)** *Skull of Small Kangaroo Found at Riversleigh*, 2010, Australian Heritage Photographic Library, Department of the Environment and Energy, rt56502, courtesy Department of the Environment and Energy; **156–157** Peter Schouten, *Reconstruction of Diprotodon optatum*, page 63 in *Prehistoric Animals of Australia* by Susan Quirk and Michael Archer (eds) (Sydney: The Australian Museum, 1983), nla.cat-vn863878, courtesy Peter Schouten; **158** *KKCC Site, Riversleigh*, courtesy Robert Lohr; **159** *Fossil in Limestone at Riversleigh*, 2009, Australian Heritage Photographic Library, Department of the Environment and Energy, rt56500, courtesy Department of the Environment and Energy; **160** Dragi Markovic, *Entering the Main Hall of the Cave, Naracoorte National Park*, 2007, Department of the Environment and Energy, eadig52396, courtesy Department of the Environment and Energy; **161** *Skull of the Giant Leaf Eating Kangaroo, Naracoorte*, 2010, Australian Heritage Photographic Library, Department of the Environment and Energy, rt56508, courtesy Department of the Environment and Energy; **162–163** *Bone Accumulation in the Upper Ossuary, Victoria Fossil Cave, Naracoorte*, 2010, Australian Heritage Photographic Library, Department of the Environment and Energy, rt56483, courtesy Department of the Environment and Energy; **163 (inset)** *Leaf Eating Kangaroo Fossil, Naracoorte*, 2010, Australian Heritage Photographic Library, Department of the Environment and Energy, rt56464, courtesy Department of the Environment and Energy; **164** Bob Nicol (photographer), Australian Information Service, *Dr Archer with a Partial Skull of a Species of Euryzygoma, Ayr, Queensland*, 1975, nla.cat-vn7578415; **165** Tony Walters, *Phil Creaser, Lizard Cannell and Chris Larkin Prepare the Rock for Explosives To Be Placed Inside to Break it Apart, Wholly Dooley, Outside the World Heritage Area Riversleigh in Far North West Queensland*, 2013, Fairfax Syndication, FXT174774, fairfaxsyndication.com.

MACQUARIE ISLAND

166–167 Peter Dombrovskis (1945–1996), *Royal Penguin Rookery, Macquarie Island, 1984, 6*, nla.cat-vn5018010; **168** Peter Dombrovskis (1945–1996), *Boulders near Summit of Mount Hamilton, Macquarie Island, 1984* (detail), nla.cat-vn4938810; **170** Peter Dombrovskis (1945–1996), *Basking Elephant Seals, Macquarie Island, 1984*, nla.cat-vn5976321; **171** Graham Robertson, *Wandering Albatross Courting*, 1999, courtesy Graham Robertson; **172** Peter Dombrovskis (1945–1996), *King Penguins near Sandy Bay, Macquarie Island, Tasmania*, 1984, nla.cat-vn2204444; **174** Peter Dombrovskis (1945–1996), *Mount Kidson, Macquarie Island, Tasmania, 1984, 1* (detail), nla.cat-vn4974815; **175** Felicity Jenkins, *Historical Artefacts of Past Seal Hunters Lay in an Outdoor Museum at Macquarie Island, Tasmania, 17 January 1998* (detail), nla.cat-vn796834; **176** Frank Hurley (1885–1962), *Sir Douglas Mawson Standing by a Whale Boat on the Beach at Caroline Cove, Macquarie Island, Australasian Antarctic Expedition*, between 1911 and 1914, nla.cat-vn1614868; **177 (top)** Laurence Le Guay (1917–1190, photographer), Australian Information Service, *Examining a Royal Penguin, Macquarie Island, 1948*, nla.cat-vn7762164; **(bottom)** Felicity Jenkins, *A Fur Seal Is Captured and Harnessed To Be Weighed and Tagged at a Colony at Secluded Bay, Macquarie Island, Tasmania*, 1998, nla.cat-vn2319276; **178–179** BMJ, *King Penguins, Macquarie Islands, Australia*, Shutterstock, 111279722.

HEARD AND MCDONALD ISLANDS

180–181 Laikolosse, *Morning Light on Fresh Ice, Heard Island*, 2016, flickr.com/photos/laikolosse/33211610323, reproduced under CC BY 2.0; **182–183** Laikolosse, *Big Ben* (detail), 2016, flickr.com/photos/laikolosse/33636744011, reproduced under CC BY 2.0; **184** Laikolosse, *Macaroni Penguin (Eudyptes chrysolophus)* (detail), 2016, flickr.com/photos/laikolosse/33180804104, reproduced under CC BY 2.0; **185** National Aeronautics and Space Administration Earth Observatory, *Lava Flow on Mawson Peak, Heard Island*, 2013, GSFC_20171208_Archive_e001509; **186–187** Frank Hurley (1885–1962), *Elephant Seal Pups in a Secluded Corner of Atlas Cover, Heard Island, Antarctica*, c.1930, nla.cat-vn2303085; **188** Ed Dunens, *Black-browed Albatross* (detail), 2015, flickr.com/photos/blachswan/19891347002, reproduced under CC BY 2.0; **190** Frank Hurley (1885–1962), *The Hut in Which Eight Scientists Sheltered in November 1929 during a Prolonged Blizzard, Heard Island, Antarctica*, c.1930, nla.cat-vn2302829; **191** Australian News and Information Bureau, *Australians Set Up World's Loneliest Weather Post, Antarctica*, 1948, nla.cat-vn7762136; **192** Doug Thost, *Botanists Jenny Winham and Patricia Selkirt Collecting Plant Samples on Heard Island, Antarctica, October 2000*, nla.cat-vn4981347; **193** Doug Thost, *Light Amphibious Resupply Cargo Vessel Transporting Gear with Stephenson Glacier in Background, Heard Island, Antarctica, February 2004*, nla.cat-vn4981335.

GREATER BLUE MOUNTAINS AREA

194–195 Andrii Slonchak, *Sunrise in Blue Mountains. View on Three Sisters Rock Formation*, Shutterstock, 359155208; **196** Emmanual Conomos, *Wentworth Falls* (detail), p.487 in *Blue Mountains Geographical Encyclopaedia* by Brian Fox, Michael Keats and John Fox (Sydney: Keats Holdings Pty Ltd, 2018), nla.cat-vn7501187, courtesy Emmanual Conomos; **198** Stephanie Morris, *Eucalypts, Blue Mountains National Park* (detail), 2018, courtesy Stephanie Morris; **199** Nick Moir (b.1975), *A Raging Bushfire in the Grose Valley, Blue Mountains, New South Wales, 21 November 2006*, nla.cat-vn4595114; **200–201** Greg Bourke, *Wollemi Pine (Wollemia nobilis)* (detail), 2018, courtesy Greg Bourke; **202** Jamie C., *Red Hands Cave, Blue Mountains*, 2013, flickr.com/photos/jamiec2009/11606763176, reproduced under CC BY 2.0; **203** Robert Wendel (artist) and Charles Troedel & Co. (lithographer), *Zig-Zag Railway, Blue Mountains* (detail), 1878, nla.cat-vn2666463; **204–205** courtesy Mike Elliott; **206** Kerry & Co., *Myles Dunphy and Bert Gallop at Jenolan Caves* (detail), 1912, State Library of New South Wales, FL3312011.

PURNULULU NATIONAL PARK

208–209 Robyn Jay, *Purnululu* (detail), 2014, flickr.com/photos/learnscope/15756330118, reproduced under CC BY-SA 2.0; **210** Richard Woldendorp (b.1927), *Bungle Bungle National Park, East Kimberley, Western Australia, near the Northern Territory Border* (detail), *1985*, nla.cat-vn3102342; **211** Philip Schubert, *Vertical View of Livistonia Palms Reaching Skywards in the Entrance of Echidna Chasm in the Bungle Bungles World Heritage Listed Purnululu National Park, Western Australia* (detail), Shutterstock, 633759566; **212–213** Steve Strike (photographer), Australian Pacific Touring Pty Ltd, *Cathedral Gorge, Bungle Bungles, Purnululu National Park, The Kimberley, WA*, 2016, Tourism Australia, 647278; **214 (top)** Sara Winter, *The Kungkalahayi Lookout in Purnululu National Park*, Shutterstock, 314371502; **(bottom)** Peter Valentine, *Sandstone Shrike Thrush, Purnululu*, 2011; **216** Philip Schubert, *Aerial of the Bungle Bungles during the Wet Season*, Shutterstock, 513221149.

ROYAL EXHIBITION BUILDING AND CARLTON GARDENS

218–219 Michelle McAulay, *Ceiling of the Royal Exhibition Building*, 2008, Department of the Environment and Energy, digo05972, courtesy Department of the Environment and Energy;
220–221 Greg Hahn, *Reflecting on Carlton Gardens*, 2010, flickr.com/

photos/atilla2008/4631634018, courtesy Greg Hahn; **222** J. Batten (printer), *Exterior of the Melbourne International Exhibition*, c.1880, nla.cat-vn359822; **223** J. Batten (printer), *The Melbourne International Exhibition 1880: Plan of Gardens*, c.1880, nla.cat-vn666394; **224** *Victorian Locomotives, Melbourne International Exhibition*, 1880, nla.cat-vn2189030; **225** Herbert Henry Fishwick (1882–1957, photographer), *Fountain at the Front Entrance of the Exhibition Building, Melbourne* (detail), 1933, nla.cat-vn6329166, courtesy Fairfax Syndication, fairfaxsyndication.com; **226–227** Greg Power (b.1972), *Royal Exhibition Building, Carlton Gardens, Melbourne* (detail), 2007, nla.cat-vn4391331; **228** Tom Roberts (artist), Australian Art Association, *Opening of the First Parliament of the Commonwealth, 9th May 1901*, 1903, nla.cat-vn1188679; **229** Dragi Markovic, *Royal Exhibition Building and Carlton Gardens, Victoria St, Carlton*, 2008, Department of the Environment and Energy, dig011749, courtesy Department of the Environment and Energy.

SYDNEY OPERA HOUSE

230–231 Druid007, *Sydney Opera House in the Evening on January 9, 2019*, Shutterstock, 1283632906; **232** Don McMurdo (1930–2001), *Harbour-side Foyer of the Sydney Opera House Concert Hall*, between 1976 and 1998, nla.cat-vn11749; **233** David Moore (1927–2003), *Sydney Opera House, 1968*, nla.cat-vn736152; **234 (top)** Jørn Utzon (1918–2008), *East Elevation*, plate in *Sydney National Opera House* by Jørn Utzon (Copenhagen: Atelier Elektra: 1958), nla.cat-vn2682346, courtesy Jan Utzon; **(bottom)** Jørn Utzon (1918–2008), *North Elevation*, plate in *Sydney National Opera House* by Jørn Utzon (Copenhagen: Atelier Elektra: 1958), nla.cat-vn2682346, courtesy Jan Utzon; **235** Jørn Utzon (1918–2008), *Site Plan*, plate in Sydney *National Opera House* by Jørn Utzon (Copenhagen: Atelier Elektra: 1958), nla.cat-vn2682346, courtesy Jan Utzon; **236** Wolfgang Sievers (1913–2007), *View from the Stage of the Sydney Opera House*, c.1973, nla.cat-vn3296054; **238–239** Pisa Photography, *Aerial View of Sydney Harbor and Downtown Skyline, Australia*, Shutterstock, 1121367998; **240** Australian Information Service, *The Queen in Sydney for an Opening at the Opera House, New South Wales*, 1973, National Archives of Australia, A6180, 1/11/73/49; **241** John Coburn (artist), New South Wales Department of Tourism, *Sydney Opera House Festival, Curtain Rises 20th October*, 1973, courtesy Sydney Opera House Trust and the John Coburn Family Estate, © John Coburn/Copyright Agency, 2018, nla.cat-vn1899747.

AUSTRALIAN CONVICT SITES

242–243 Wolfgang Sievers (1913–2007), *Church and View from Scorpion Hill, Port Arthur, Tasmania* (detail), 1965, nla.cat-vn3411269; **244–245** *Cascades Female Factory, from the East, South Hobart* (detail), 1900, Tasmanian Archive and Heritage Office, NS1013/1/48; **246** Dragi Markovic, *Brickendon Estate* (detail), 2007, Department of the Environment and Energy, dig010094, courtesy Department of the Environment and Energy; **247** David Lochlin, *Port Arthur Tasmania Panorama* (detail), 2008, flickr.com/photos/dlochlin/9552178006, reproduced under CC BY-SA 2.0: creativecommons.org/licenses/by-sa/2.0; **248** *Entrance of the Machine Shop on Cockatoo Island, New South Wales* (detail), 1927, nla.cat-vn6264813, courtesy Fairfax Syndication, fairfaxsyndication.com; **250–251** Dragi Markovic, *Main Cell Block, Fremantle Prison (Former)*, 2007, Department of the Environment and Energy, dig011186, courtesy Department of the Environment and Energy; **252** William Pedersen, Australian News and Information Service, *All That Remains of the Treadmill, One of the Merciless Punishments Inflicted on Convicts during Norfolk Island's Era as a Penal Settlement for Hardened Offenders* (detail), 1953, nla.cat-vn4590004; **253** William Pedersen, Australian News and Information Bureau, *Gallows Gate, All That Remains of the Grim Prison of the Convict Era on Norfolk Island*, 1953, nla.cat-vn4589994.

NINGALOO COAST

254–255 Matt Curnock, *Ningaloo Reef* (detail), 2018, courtesy Matt Curnock; **256–257** Darkydoors, *Aerial View of Cape Range National Park and the Ningaloo Marine Park, Exmouth, Western Australia* (detail), Shutterstock, 770005099; **258–259** Matt Curnock, *Ningaloo Reef with Fish* (detail), 2018, courtesy Matt Curnock; **260** Jane Ambrose, *Yardie Creek* (detail), 2008, Department of the Environment and Energy, dig005680, courtesy Department of the Environment and Energy; **261** ChGross1, *Pink Flowers (Ipomea yardiensis) near Yardie Gorge* (detail), 2014, flickr.com/photos/chgross/34957758094, reproduced under CC BY-SA 2.0; **262–263** Matt Curnock, *Whale Shark, Ningaloo* (detail), 2018, courtesy Matt Curnock; **264–265** Julien Carnot, *Yardie Creek*, 2011, flickr.com/photos/julien-carnot/6159218995, reproduced under CC BY-SA 2.0.

BUDJ BIM CULTURAL LANDSCAPE

266–267 Tyson Lovett-Murray, *Drone Image of Weir*, Gunditj Mirring Traditional Owners Aboriginal Corporation; **268** Tyson Lovett-Murray, *Aquaculture Channel and Pond*, Gunditj Mirring Traditional Owners Aboriginal Corporation; **269** Justin McManus, *Tyson Lovett-Murray Holding Kooyang*, Gunditj Mirring Traditional Owners Aboriginal Corporation; **270–271** Cafuego, *Lake Surprise @ Budj Bim*, 2018, flickr.com/photos/cafuego/42824845171, reproduced under CC BY 2.0; **272** Denisbin, *Lake Condah. Remains of 1,700 Year Old Aborginal Stone House*, 2015, flickr.com/photos/82134796@N03/17437595495, reproduced under CC BY 2.0; **274** Tyson Lovett-Murray, *Lake Condah*, Gunditj Mirring Traditional Owners Aboriginal Corporation; **275 left** Tyson Lovett-Murray, *Channel with Reconstructed Stake, Branch and Weir*, Gunditj Mirring Traditional Owners Aboriginal Corporation; **275 right** Denisbin, *Woven Eel Trap*, 2015, flickr.com/photos/82134796@N03/17570619300, reproduced under CC BY 2.0; **276** Cafuego, *Swamp Wallaby*, 2018, flickr.com/photos/cafuego/41014360800, reproduced under CC BY 2.0; **277** Tyson Lovett-Murray, *Tae Rak Visitor Experience Area: Jetty and Boat Ramp*, Gunditj Mirring Traditional Owners Aboriginal Corporation.

POTENTIAL NEW WORLD HERITAGE SITES IN AUSTRALIA

278–279 Trevern Dawes, *Sand Dunes at Lake Eyre North, South Australia*, 2009, nla.cat-vn6333472; **281** Peter Valentine, *Australian Heritage Council at Murujuga*, 2010; **282–283** Trevern Dawes (b.1944), *Sand Dunes at Lake Eyre North, South Australia*, 2009, nla.cat-vn6333472.

THE FUTURE OF WORLD HERITAGE IN AUSTRALIA

284–285 Kerry Trapnell, *Weaving at Laura Dance Festival* (detail), 2005, courtesy Kerry Trapnell; **286** Maurizio Leonardi, *Mitchell Falls in North West Western Australia*, Shutterstock, 1464802322; **288** Peter Valentine, *Green Python, Adult, Cape York Peninsula*, 2009.

ENDMATTER

292–293 Alex Couto, *Purnululu National Park*, Shutterstock, 559335664; **302–303** Janelle Lugge, *King Penguins, Aptenodytes patagonicus* on Macquarie Island, Sub-antarctic Islands, Australia, 58273180; **312** *Professor Peter Valentine* © Martin Willis, **312–313** Peter Valentine, *Wompoo Fruit-dove in a Fig Tree*.

Index

Page numbers in **bold** indicate illustrations.

Peter Valentine is an adjunct professor at James Cook University, following an academic career in environmental management and conservation based in Townsville. He has written numerous peer-reviewed articles, book chapters and monographs on this subject. His focus on matters World Heritage began through involvement in Australian tropical rainforest conservation and he subsequently became a member of the IUCN World Commission on Protected Areas in 1989, a position that he still holds. For several years, he was the Commission's Vice Chair for Publications. He spent sabbatical time at the IUCN's headquarters in Switzerland as a World Heritage Fellow in 1995, working with Dr Jim Thorsell on World Heritage nomination and evaluation processes and the role of the IUCN. He has been intensely involved in natural heritage across Australia and throughout the world for many decades, assisting with the evaluation of nominations; establishing a graduate university program in World Heritage management; participating in the management and evaluation of several World Heritage sites; serving terms as director and chair of the Wet Tropics Management Authority; and serving as a natural heritage expert on the Australian Heritage Council. He has visited most Australian World Heritage areas and many more World Heritage sites overseas. Today, he lives with, wife, Valerie near Malanda in the Wet Tropics rainforest.

A wompoo fruit-dove, a common if elusive resident of the Wet Tropics.

NLA The publishing imprint of the National Library of Australia

Printed by National Library of Australia Publishing
Canberra ACT 2600

ISBN: 9781922507587

First published 2019. Second edition published 2023.

First Nations people are advised this book contains depictions and names of deceased people, and content that may be considered culturally sensitive. Every reasonable endeavour has been made to correctly name Traditional Owners and Custodians. The publisher welcomes any corrections or additions to this information; please contact the Library.

Books published by the National Library of Australia further the Library's objectives to produce publications that interpret the Library's collection and contribute to the vitality of Australian culture and history.

Commissioning Publishers: Susan Hall, Lauren Smith
Managing editor: Amelia Hartney
Editor: Katherine Crane
Designer: Amy Cullen, Mike Ellott
Image coordinators: Stephanie Morris, Jemma Posch
Indexer: Sherrey Quinn
Printed by C&C Offset Printing Co., Ltd on FSC®-certified paper.

Find out more about NLA Publishing at publishing.nla.gov.au.

NATIONAL LIBRARY OF AUSTRALIA

A catalogue record for this book is available from the National Library of Australia

FSC www.fsc.org MIX Paper | Supporting responsible forestry FSC® C008047